군사 영어

한종철 저

형설출판사
www.hyungseul.co.kr

머리말 PREFACE

본 교재는 부사관과 학생으로서 일상생활 및 군 생활에 필요한 기본적인 실용영어 능력을 향상하는데 기본적인 목적을 두고 있다. 부사관으로서 영어능력전반을 향상시키기 위해 기본문법과 군 관련 어휘를 활용한 생활영어 학습을 통해 국제화 시대의 엘리트 군 지휘관으로서 갖춰야할 기본적 영어감각을 향상시키는데 있다.

- 군사영어 I 에서는 기본적인 군사용어를 학습하게 된다.
- 군사영어 II 에서는 국군 방송에서 방영했던 군사영어 가운데 실생활 및 군 생활에서 활용 가능한 실용회화를 학습하게 된다.
- 실용영어에서는 기초문법학습과 더불어 공군, 해군, 육군3사관 지원생들을 위한 영어평가 대비학습을 하게 된다.

군의 중추역할을 하는 부사관은 스스로 명예심을 추구하여 빛남으로 자긍심을 갖게 되고, 사회적인 인간으로서 지켜야 할 도리를 지각하면서 행동할 수 있어야 하며, 개인보다는 상대를 배려할 줄 아는 공동체 의식을 견지하며 매사 올바른 사고와 판단으로 건설적인 제안을 함으로서 내가 속한 부대와 군에 기여하는 전문성을 겸비한 인재들이다. 21세기 떠오르는 직업으로 각광받고 있는 부사관은 국가 공무원으로써의 신분을 가지게 됩니다. 폭넓은 시각과 심화된 전문성을 겸비한 인재가 되기 위해 본 책의 학습을 통해 한걸음 나아가길 바랍니다.

목 차

PREFACE

제1부 군사영어 | • 8

군사영어 Ⅱ • 32

제3부 실용영어(Practical English) • 60

Military English

군사영어 I

1 부대 단위(Military Unit)

<table>
<tr><th colspan="2">육군</th><th colspan="2">해군</th><th colspan="2">공군</th></tr>
<tr><td>군</td><td>Army</td><td>함대</td><td>Fleet</td><td>방공포
사령부</td><td>Air
Defence
Artillery
Command</td></tr>
<tr><td>군단</td><td>Corps</td><td>전단</td><td>Flotilla</td><td>전투비행단</td><td>Fighter Wing</td></tr>
<tr><td>사단</td><td>Division</td><td>전대</td><td>Squadron</td><td>전대</td><td>Squadron</td></tr>
<tr><td colspan="3">여단</td><td colspan="3">Brigade</td></tr>
<tr><td colspan="3">연대</td><td colspan="3">Regiment</td></tr>
<tr><td colspan="3">대대</td><td colspan="3">Battalion</td></tr>
<tr><td colspan="3">중대</td><td colspan="3">Company</td></tr>
<tr><td colspan="3">소대</td><td colspan="3">Platoon</td></tr>
<tr><td colspan="3">분대</td><td colspan="3">Section/Squad</td></tr>
</table>

• section/squad (분대 : 9~13soldiers)	In military terminology, a squad is a small military unit led by a non-commissioned officer (NCO) that is subordinate to an infantry platoon.
• platoon (소대 : 20~50soldiers)	A platoon is a military unit typically composed of two to four sections or squads and containing 25 to 50 soldiers.
• company (보병 중대 : 70~200soldiers)	A company is a military unit, typically consisting of 75~200 soldiers and usually commanded by a Captain or a Major. Most companies are formed of three to five platoons.
• battalion (일반 대대 : 600~1500soldiers)	A battalion is a military unit of around 600~1500 soldiers usually consisting of between two and seven companies and typically commanded by either a Lieutenant Colonel or a Colonel. Several battalions are grouped to form a regiment or a brigade.
• regiment (연대) (소속의 대대가 같은 병과인 경우)	A regiment is a military unit, composed of variable numbers of battalions, commanded by a Colonel.
• brigade (여단 : 2000~7000soldiers) (소속의 대대가 다른 병과인 경우)	A brigade is a military formation that is typically composed of two to five regiments or battalions, depending on the era and nationality of a given army. Usually, a brigade is a sub-component of a division, a larger unit consisting of two or more brigades.
• division (사단 : 10000~15000soldiers)	A division is a large military unit or formation usually consisting of between 10,000 and 30,000 soldiers. In most armies, a division is composed of several regiments or brigades, and in turn several divisions make up a corps.
• corps (군단 : 20000~45000soldiers)	In many armies, a corps is a battlefield formation composed of two or more divisions, and typically commanded by a lieutenant general.
• army	army/navy/air force

군 부대의 단위

| 군 부대의 단위를 지칭하는 단어들을 활용한 독해지문을 읽고 해석해 보세요.

Listen! Our Squad Shares A Single Fate, We Shall Move Together At All Costs.

「squad는 병사 10명과 하사 2명으로 구성된 '분대'를 뜻합니다. 미군에서 사용되는 용어입니다. 영국군에서는 section이라는 표현이 '분대'를 의미하며 이 단어 section이 미군에서 사용될 경우 '반(半) 분대'를 뜻합니다.」

Helicopters thrummed in to deposit entire platoons of infantrymen, and armored personnel carriers rumbled through the mire at 450 yards an hour in hopes of pinning the enemy troop to the sticking point.

「platoon은 '소대'입니다. 2개 분대 이상으로 편성된 군 단위입니다. 규모는 30명 정도입니다.」

There must be more reinforcements to keep the isolated company from annihilation.

「company는 4개의 소대로 편성되어 대위(大尉)가 지휘하는 '중대'입니다. 보통 120명의 군사가 있습니다. 참고로 보병 중대는 an infantry company, 기병 중대는 a troop, 포병 중대는 a battery, 비행 중대는 영국의 경우 a squadron이라 합니다.」

The fifth battalion will go to Iraq to supplement the losses.

「battalion은 '대대'를 뜻합니다. 보통 500에서 1000명으로 구성된 꽤 큰 규모의 부대입니다.」

➡ The regiment suffered a deterioration of discipline as the newly employed colonel displayed serious inconsistency in his leadership.

「regiment는 '연대'입니다. 보통 연대는 3개의 보병 대대로 이루어져 있습니다.」

➡ Under the United Nations flag, the British brigade, including Australian, Canadian and New Zealand troops, participated in the Kapyong battle against North Koreans backed by Chinese forces in 1951.

「brigade는 보통 2개 이상의 연대로 이루어진 '여단'입니다. 여단은 보통 장성에 의해 지휘됩니다. 마지막으로 이 여단이 2개 이상 모여 20,000명 규모의 division(사단)이 되고, 다시 이 division(사단)이 2개 이상 모여 corps(군단)을 이룹니다.」

2 병과(Branches)

• infantry(보병)	Infantry is the branch of a military force that fights on foot—troops specifically trained to engage, fight, and defeat the enemy in face-to-face combat.
• cavalry(armor)(기갑)	The cavalry is the part of an army that uses armoured vehicles for fighting.
• artillery(포병)	Artillery consists of large, powerful guns which are transported on wheels and used by an army.
• engineer corps(공병)	Combat engineer is an individual soldier who performs a variety of combat engineering duties. Such tasks typically include bridge-building, laying or clearing minefields, demolitions, field defences, and building, road and airfield construction and repair.
• aviation, air combat(항공)	Aviation is the operation and production of aircraft.
• quartermaster(보급)	It is a term referring to either an individual soldier or a unit, who specializes in distributing supplies and provisions to troops.
• ordnance(병기)	Military supplies including weapons, ammunition, combat vehicles, and maintenance tools and equipment.
• maintenance(정비)	The maintenance of a building, vehicle, road, or machine is the process of keeping it in good condition by regularly checking it and repairing it when necessary.
• transportation(수송)	Transportation refers to any type of vehicle that you can travel in or carry goods in.
• medic(의무)	Medic is a general term for a person involved in medicine, especially emergency or first-response medicine, such as an emergency medical technician, paramedic, or a military member trained in battlefield medicine.

• logistics(병참, 군수)	In military logistics, logistics officers manage how and when to move resources to the places they are needed.
• special forces(특전)	Special operations forces are military units trained to perform unconventional missions.[1] Special forces, as they would now be recognised, emerged in the early 20th century, with a significant growth in the field during the Second World War.
• military intelligence(정보)	Military intelligence is a military service that uses intelligence gathering disciplines to collect informations which may be of use in the decision making process of commanders.
• military police(헌병)	Military police(MPs) are normally the police of a military organization.
• communications/signal(통신)	Communication is a process of transferring information from one entity to another.
• rangers(특공)	A specially trained elite unit of the United States Army.
• air borne infantry(공수)	Airborne troops could be dropped on the battlefield either via parachute or in gliders.
• CBR	chemical, biological, and radiological
• air assault infantry	Air assault is the movement of ground-based military forces by vertical take-off and landing (VTOL) aircraft—such as the helicopter—to seize and hold key terrain which has not been fully secured, and to directly engage enemy forces.

Amy Branch Marks

*출처 : 대한민국 육군(www.army.mil.kr)

Distintion	Branch Marks	Branch Mascot	Meaning
Infantry			–Represented with sword and shield –The basic weapon for ancient warriors
Artillery			–Two gun ship in cross formation, train, and the head cover represent artillery.
Armor			–Since the emerging of tanks after WWI, there has been changes made to cavalry's role, and thus included the form of tank to represent the development from cavalry to armor. –Hoof is added to the infantry logo to represent cavalry.
Intelligence			–A bat represents intelligence activities, the eye repres ents enemy territory observation, the talon represents collection and evaluation, and the wings represent the universe. –The logo represents the development from the usage of torch in the ancient time to the usage of night binocular to observe the enemy and their information collection.
Engineer	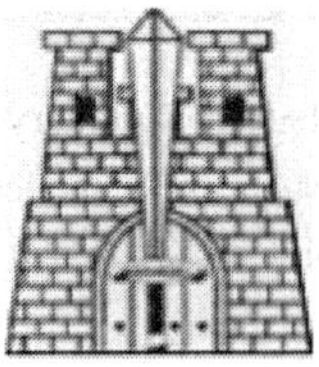		–Independence gate represents construction, and the star represents the nations' independence and protection. –Korea's typical ancient fortress and sword represents combat engineersollection.

Signal			-Two crossed hand signal flags and a torch in the middle represent this branch. -Modified signal fire and the wings of doves. -The ancient means of communication to the modern microwave communication equipment is described.
Aviation			-The wings represent the air, and shield, sword, and arrow represent the fire power support.
Chemical			Benzene핵의 분자식을 구조적으로 간결하게 표시하였으며 여러 화학 약품과 장비 표시를 통해 군화학 임무의 중요성을 강조
Weapon			고대병기인 활과 화살 및 죽화통을 통해 경과의 특징을 표시
Logistics			병참 열쇠는 보급통제기능, 펴잎은 1종 보급품, 날개는 보급의 신속성을 상징
Transportation			배는 차륜속에 방패가 들어있고, 방패속에는 날개를 단 차륜이 철로 위에 놓여있는 형태를 통해 5대 수송 수단인 육로, 수로, 선로, 철로, 송유관의 수송기능을 상징

3 계급(Ranks)

■ 병(Enlisted man)

병계급장의 형태는 지구 구성요소인 4개의 층을 표시한 것으로 군의 기반 형성의 상징으로 계급이 오를 수록 전투 능력이 향상 및 임무수행의 숙달을 의미합니다.

계급	명칭	계급장	설명
이병	• Private • Seaman Apprentice • Airman		
일병	• Private first class • Seaman • Airman First Class		When South Korea's national defense guard was founded, enlisted ranks were divided in to private and private first class. These terms have been used ever since. 1946년 남조선 국방 경비대 창설시 병 계급은 2등급으로 이등(二等), 일등(一等) 개념으로 구분된 것이 현재까지 명칭되어 지고 있습니다. When South Korea's national defense guard was founded, enlisted ranks were divided in to private and private first class. These terms have been used ever since.
상병	• Corporal • Petty Officer Third Class • Senior Airman(sergeant)		In 1962, enlisted ranks were divided into four, and those who were in upper rank of the enlisted were called corporal. 1962년 병계급이 4등급 체제로 바뀌자 병 계급 중 상위 계급이라 하여 상등(上等)병이라 칭하게 되었습니다.
병장	• Sergeant • Petty Officer Second Class • Staff Sergeant		The term sergeant was used to signify the head of enlisted ranks. 병들중에서 우두머리에 속하는 계급이므로 우두머리 장(長)을 사용하여 병장이라고 칭하게 되었습니다.

■ 부사관(Non–commissioned officer)

In 1996, Non–commissioned officers' insignia was revised from the past insignia on which their insignia was marked on top of that of the enlisted soldiers causing it to look too big. So to arouse their pride and to make it look like the insignia of the officers, they added the Mugunghwa sign underneath their insignia.

<table>
<tr><td>하사</td><td>• Staff Sergeant
• Petty Officer First Class
• Technical Sergeant</td><td></td><td rowspan="4">1996년 부사관 계급장이 개정되었는데, 그 이전에는 병장 계급장위에 부사관 계급을 표시하였으나 크기가 크고 간부로서 위상제고를 위해 장교와 유사한 형태로 개정하였으며, 굳건한 기초위에 자라나는 나뭇가지를 형상화 한 것으로 자라나는 나무처럼 전문화된 기술, 숙련된 전투력 개발능력의 축적은 나타내었습니다.</td></tr>
<tr><td>중사</td><td>• Sergeant First Class
• Chief Petty Officer
• Master Sergeant</td><td></td></tr>
<tr><td>상사</td><td>• Master Sergeant
• Senior Chief Petty Officer
• Senior Master Sergeant</td><td></td></tr>
<tr><td>원사</td><td>• Sergeant Major
• Master Chief Petty Officer
• Chief Master Sergeant</td><td></td></tr>
</table>

■ 위관장교(Company grade officer)

Company officer's rhombus symbolizes the beginning officers' strong will of the protection of the nation and is expressed in a form of strong and unbreakable diamond

<table>
<tr><td>준위</td><td>• Warrant Officer</td><td> </td><td rowspan="4">위관장교의 마름모의 금강석(다이아몬드)을 상징하는 것으로 가장 단단하면서 깨어지지 않는 특성을 초급장교로서 국가 수호의 굳건한 의지로 표현하였습니다.</td></tr>
<tr><td>소위</td><td>• Second Lieutenant
• Ensign</td><td> </td></tr>
<tr><td>중위</td><td>• First Lieutenant
• Lieutenant Junior Grade</td><td> </td></tr>
<tr><td>대위</td><td>• Captain
• Lieutenant</td><td> </td></tr>
</table>

■ 영관장교 및 장군(Field grade and general officer)

Field officers' bamboo symbolizes the ever greenness of four seasons, strong spirit, and fidelity

소령	• Major		영관의 대나무는 사계절 항상 푸르름과 굳건한 기상 그리고 절개를 상징합니다.
중령	• Lieutenant Colonel • Commander		
대령	• Colonel • Captain		

준장	• Brigadier General • Rear Admiral(Lower Half)		장군의 별은 스스로 빛을 내는 천체로서 군에서의 모든 경륜을 익힌 완숙한 존재임을 상징하며 위관장교의 금강석은 지하를 영관장교는 시상을 장군의 별은 우주를 표현합니다.
소장	• Major General • Rear Admiral(Upper Half)		
중장	• Lieutenant General • Vice Admiral		
대장	• General • Admiral		

4 직책(Positions)

참모총장	Chief of Staff
사령관	Army Commanding General
군단장	Corps Commanding General
사단장	Division Commanding General
여단장	Brigade Commander
연대장	Regiment Commander
대대장	Battalion Commander
주임원사	Command Sergeant Major
중대장	Company Commander
행정보급관	First Sergeant
소대장	Platoon Leader
부소대장	Platoon Sergeant
분대장	Squad Leader

5 전술용어(Tactics Words)

대대전술훈련평가	ATT : Army Training Test
연대전술훈련평가	RCT : Regiment Combat Test
야외기동훈련	FTX : Field Training Exercise
혹한기훈련	Cold Weather Exercise(training)
임무형 보호태세	MOPP : Mission Oriented Protective Posture
지휘, 통제, 통신, 컴퓨터, 정보	C4I : Command, Control, Communications, Computer, Intelligence
임무, 적 상황, 지형/기상, 가용병력, 가용시간, 민간요소	METT-TC : Mission, Enemy Information Terrain/Weather, Troops&Support available, Time available, Civil Consideration
완전군장	Full Combat Gear
피아식별	IFF : Identification Friend or Foe
합동작전	Joint Operations
근접항공지원	Close Air Support
전쟁준비물자	War Readiness Material
전투근무지원	Combat Service Support
설영대	Quartering Party
증강	Reinforcement
역습	Counter Attack

6 군대에서 많이 쓰이는 군사 일반용어 영어표현

- 인식표 : identification tag
- 임기응변 : improvisation
- 재고현황 : stock status
- 적성검사 : aptitude test
- 전면전쟁 : general war
- 전속(轉屬) : permanent change of assignment(PCA)
- 접근로 : avenue of approach
- 정식간격 : normal interval
- 정신훈화 : edifying talk
- 제공권 : air supremacy
- 조명탄 : illuminationl shell
- 지구전 : sustained conflict
- 집결지역 : assembly area
- 집중사격 : concentrated fire
- 철거 : roll-up

* 철수 : withdraw

* 징병제도 : conscription system

* 철회 : retirement

- 총사령관 : commander-in-chief(CINC)
- 최루가스 : lachrymatory[tear] gas
- 침투부대 : infiltration force
- 탄띠 : ammunition belt
- 파편 : fragment
- 폭파 : demolition
- 피아식별장치 : identification friend or foe(IFF)
- 항명(抗命) : insubordination
- 인해전술 : human wave tactics
- 자해행위 : self-infliction
- 재물조사 : inventory
- 전령 : messenger
- 전면포위 : encirclement
- 전투식량 : C-ration
- 전투준비태세 : combat readiness conditions
- 정면공격 : frontal attack
- 정신교육 : motivational education
- 정훈 : troop information and education(TI&E)
- 제독(除毒) : decontatmination
- 즉결처분 : non-juridical punishment
- 지휘소 연습 : command post excercise(CPX)

* 기동훈련 : maneuver excercise(MX)

- 징발/징병 : requisition

* 징집병/신병 : conscript

- 총동원 : total mobilization
- 최대병력 : peak strength
- 총소요 : gross requirement
- 최루탄 : tear shell
- 취약성 : vulnerabilities
- 탄도수정 : ballistic correction
- 파견근무 : temporary duty(TDY)
- 폭발물처리 : explosive ordnance disposal
- 폭풍효과 : blast effect
- 하사관 : non-commissioned officer(NCO)
- 항복 : surrender
- 핵공격 경고 : nuclear strike warning

- 해체부대 : deactivated unit
- 핵관측 보고 : NBC(Nuclear, Biological and Chemical) Report
- 현 보유 : on hand(O/H)
- 현재원 : strength for duty
- 현지임관 : battlefield commission
- 호출부호 : call sign
- 화염방사기 : flame thrower
- 화학탄 : chemical projectile
- 훈련탄 : drill ammunition
- 휴대식량 : small detachment ration
- 휴식지역 : rest area
- 휴전 : armistice
- 전복활동 : subversion activities
- 행군간 경계 : security on the march
- 행동개시 : move out
- 현역정년 : active duty expiration period
- 현지실습 : terrain excercise
- 현지조달 : local procurement
- 화생방전 : CBR(Chemical, Bilological and Radiological) warfare
- 후퇴작전 : retrograde operation
- 휴대보급품 : accompanying supplies
- 휴식시간 : off time
- 휴양 : rest and recuperation
- 흑색선전 : black propaganda

7 군대일상 용어 영어표현

- 위병 : guard
- 위생병 : corps man/medic/aidman
- 위수부대 : garrison force
- 위장군기 : camouflage discipline
- 유도탄 : guided missile
- 유동표적 : transient target
- 유선통신 : wire communication
- 유물 : remains
- 윤형철조망 : concertina
- 융단폭격 : carpet bombing
- 응급치료 : emergency medical treatment/first aid
- 외곽경계(外廓警戒) : perimeter security
- 외래환자 : out-patient
- 요새지역 : fortified area
- 용접 : welding
- 우발공격 : accidental attack
- 우선순위 : precedence/priority
- 우회(迂回) : by pass
- 운용 : employment
- 원격조정 : remote control
- 월동준비 : winterization
- 위문 : kind inquiry
- 오염지역 : area of contimination
- 완수신호 : visual signal/arm and hand signal
- 완제품 : end item
- 왕복운행소요시간 : turn around time
- 가시거리 : visibility range
- 위병소 : guard room
- 위생점검 : sanitary inspection
- 위수지 : station post
- 위험표지 : hazard signs
- 유독성 화학 작용제 : toxic chemical agent
- 유류(油類) : petroleum oil&lubricants(P.O.L.)
- 유언비어 : sensational rumors
- 유자철조망(有刺鐵條網) : barbed wire entanglement
- 음어 : code word
- 완파(完破) : total damages/complete damages
- 외과병원 : surgical hospital

＊ MASH : Mobile Army Surgical Hospital (육군이동외과병원)

- 요새진지 : fortified position
- 우군부대 : friendly troops
- 우발사태 : contingency

＊ 비상계획 : contingency plan

- 우회기동 : turning movement
- 운행증 : trip ticket
- 원대복귀 : return
- 위력수색 : reconnaissance in force
- 오발 : misfire
- 완강한 저항 : stiff resistance
- 완전무장 : individual basic load/full pack
- 완충기 : buffer

＊ 완충장치 : shock absorber

＊ 완충지대 : buffer zone

- 각개격파 : defeat[beat] in detail
- 각개훈련 : individual training
- 견책[징계] : reprimand
- 경상자 : slightly wounded
- 계엄령 : martial law

* 계엄사령부 : martial command

- 계획지침 : planning guidance
- 고립된 지대 : pocket
- 공습 : air raid
- 공습경보계통 : air raid warning system
- 공제산 : sky line
- 관사 : government quarters
- 교대 : relief/rotation
- 가용시간 : time available
- 개인호 : fox hole
- 결전 : decisive engagement
- 계급정년 : age in grade
- 계획수립 : planning
- 계획정비 : scheduled maintenance
- 고등군법회의 : general count-martial
- 공수작전 : air lift operation
- 공습경보 : air alert
- 공용화기 : crew served weapon
- 과부족 : over/short(o/s)
- 관제등 : blackout light

8 군 시설물(Army facilities)

- 군대 식당 : D-FAC(dining facility)
- 본부 중대 : headquarter
- 공군, 해군 기지 : base
- 해외 육군 임시 기지 : camp
- 공군 군내 매점 : BX(base exchange)
- 중대 본부 : oderly room
- 육군 기지 : post
- 본토 육군 영구 기지 : fort
- 육군 군내 매점 : PX(post exchange)

9 제식(Close-Order drill)

- 예령 : command
- 차렷 : Attention
- 쉬엇 : At ease
- 우로돌앗 : RIght face
- 우로[좌로]나란히 : Dress right[left]
- 바롯 : Ready front
- 뒤로 돌앗 : About face
- 우[좌]로 3보 갓 : Three steps to the right [left]
- 제 걸음으로 갓 : -Quick time march
  -Forward march
- 앞으로 갓 : Forward march
- 뒤로 돌아 갓 : To the rear march
- 우[좌]로 봤 : Eyes right[left]
- 우향[좌향] 앞으로 갓 : By the right[left] flank march
- 동령 : execution
- 열중 쉬엇 : Parade rest
- 편히 쉬엇 : Rest
- 좌로 돌앗 : Left face
- 좁은 간격 우로[좌로] 나란히 : At close interval right[left] dress
- 우[좌]로 갓 : RIght[left] step march
- 제자리 섯 : Halt
- 제자리 걸음 : Mark time march
- 반보로 갓 : Half step march
- 뛰어 갓 : Double time march
- 줄줄이 좌[우]로 갓 : Left[right] turn march
- 반우향[좌향] 앞으로 갓 : Right[left] oblique march(비스듬 : oblique)
- 보조 맞춰 갓 : Attention! Count cadence march

* cadence : 운율, 박자, 보조

- 긴 걸음으로 : −Route step march
  −At ease march
- 보조 바꾸어 갓 : Change step march
- 번호 붙여 갓 : Count cadnece count
- 좌[우]로부터 일렬지어 갓 : Column of files from the right(left) march
- 뒤로 갓 : Back step march
- 줄줄이 반우향[좌향] 앞으로 갓 : Half right [left] march
- 번호 붙여서 : By the numbers
- 거수 경례 : Hand salute, salute
- 해산 : −Fall out
  −Dismiss
- 집합 : Fall in

10 의무, 책임 등을 나타내는 관용어구

- 예정되어 있다 : be supposed to
- 확실하게 하라 : make sure, make straight
- 지도력, 통솔력 : leadership
- 설명할 수 있다 (Formation에서 인원 확인할 때 많이 쓰임) : be accounted for
- 지휘하다, 통솔하다 : take charge, be in charge of
- 허락되었다 : be allowed to
- 책임감 : responsibility
- 책임(구체적으로 설명하고 증거를 제시할 수 있는 책임) : accountability
- 각 잡힌(이건 정말 세계 공용어다) : be squared away

11 그 밖에 많이 쓰이는 용어들

- 엉망이다 : be massed up
- 검열 : inspection
- 서둘러 : Hurry up
- 가장 많이 쓰이는 인사말 :
  What's up?
  What's going on?
- 전화를 걸다 : make a phone call, give a call
- yes 대신에 긍정의 의미로 쓰임 : hooah
- 군대를 해외로 파병하다 : deploy
- 작업 : detail
- 부사관 : NCO(non-commitioned officer)
- 지휘 체계 : chain of command
- what's your position? Over!(현위치를 알려라. 이상!)
- as soon as possible : ASAP
- 미국화된 : americanized
- 신경꺼 : Never mind
- 말을 못 알아 들었을 때 다시 묻는 말 :
  What did you say?
  Say again?
  Pardon?
  What?
- no 대신에 부정의 의미로 쓰임 : negative
- 파견된 : detached
- 군대를 급파하다 : dispatch
- 장교 : officer
- 사병 : enlisted sodier

메모

Military English

군사영어 II

Military English 1-What should I call you?

1 Today's Dialogue I

A : Hi, Private. Welcome to my Unit.

B : How are you doing?

A : …?

2 Today's Dialogue II

A : Nice to meet you, private?

B : Nice to meet you, too.
But, What should I call you?

A : Just call me love.

B : O. K.

A : Anyway, why don't you go to the company headquarters to report to the company commander and check out the barracks.
I will guide you around the camp. Later.

Today's title

key express

- What should I call you?

- Just call me ___________.

- Nice to meet you.

  = Glad to meet you.

  = Pleased to meet you.

Relay Quiz

반합?

Military English 2-How was your training?

3 Today's Dialogue I

A : Come in.

B : Sir, private Byungdon Suh is reporting.

A : Don't be nervous. Take it easy.
Nice to meet you, private Suh. I'm Captain Gonzales.

B : Nice to meet you, Captain Gonzales.

A : Captain? Are you Navy?

Captain 1. 선장 2. ______

4 Today's Dialogue II

A : Nice to meet you, private Suh. I'm Captain Gonzales.

B : Nice to meet you, Captain Gonzales.

A : How was your training?

B : It was not bad. Captain Gonzales.

A : Very good. Anyway, Don't hesitate to come
and see me if you have any problem.

B : Thank you, Captain Gonzales.

A : 'Yes Ma'am' is fine.

Today's title

key express

- How was your training?
- How was the ________?

- Corporal : [해군]petty officer third class [공군]sergeant
- Sergeant :
- General :

- vacation/furlough
- guard duty

Relay Quiz

철모?
수통?

Military English 3-They move smartly and orderly.

5 Today's Dialogue I

A : What's up, private?

B : I am just watching, sir.

A : Carry on. Is it interesting?
Carry on. private.

attention :

carry on :

6 Today's Dialogue II

A : Carry on. Is it interesting?

B : Yes, It's interesting. They move smartly and orderly.

A : Your attention's pretty good.
But you should practice your hand salute a little bit more.

B : Should I?

A : No, I was just kidding.

Today's title

key express

- They move smartly and orderly.

- hand salute :

- Han Teacher speaks English __________ and __________.
  cf. fluently

- You should practice your hand salute a little bit more.
  You should practice __________ a little bit more.

- 좌향좌

- 우향우

- 뒤로 돌아

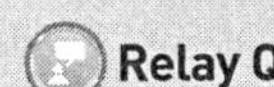

Relay Quiz

점호?

Military English 4-I will be sure to hand it over to my company commander.

7 Today's Dialogue I

A : Private Suh, do you have a moment?

B : Yes, sir.

A : Can you spare me some time, Private?

B : Sure, I am free now. What can I do for you?

A : Please take this document over to Alpha company commander.

B : No problem. Don't worry about it, sir.

sir :

8 Today's Dialogue II

A : Thank you so much, Private Suh.

B : I will be sure to hand it over to my company commander.
Is that all?

A : Just tell him that I will call him later.

B : I will.

A : See you, Private Suh. Bye.

B : I'll be seeing you, sergeant!

sergeant :

Today's title

key express

- I will be sure to hand it over to my company commander.

- hand over :

- ___________ close the window before you leave.

- _______________ my girlfriend will come to see me next week.

- Is that all?

Relay Quiz

전투화?

Military English 5-I wanted to work for the Military Police.

9 Today's Dialogue I

A : How's it going with your Combat service support assignment?

B : I am getting used to it little by little.

A : At first, I wanted to work for the Military Police.
What about you?

B : police?

military police :
a military policeman :

10 Today's Dialogue II

A : At first, I wanted to work for the Military Police.
What about you?

B : I wanted to be assigned to a tank unit.

A : Any special reasons?

B : Nothing special. I just wanted to drive a tank.

A : Why?

B : Just because tanks look so cool.
Besides, I love to drive any vehicles.

A : Anyhow, if you are driving a tank, I will not get on it.

Today's title

key express

- I wanted to work for the Military Police.
- work for

- ______________________ an infantry company.
- ______________________ a chaplain corps.
- infantry
- chaplain corps

 **Relay Quiz**

PX?

Military English 6-They say the rifle is the second life of a soldier.

11 Today's Dialogue I

A : Did you get your rifle yesterday?

B : Yes, sergeant. All the platoon got their own.

A : Very good. Now let me know how many magazines platoon has.

B : Right away?

A : Yes. It's very urgent.

B : Yes, sergeant.

magazine : 1. 2.

12 Today's Dialogue II

A : Two magazines per man. Twenty four total.

B : I issued just the same magazine to your platoon.
Did you count them yourself?

A : Yes, sergeant.

B : They say the rifle is the second life of a soldier.
Keep it in good shape.

A : Yes, I will.

Today's title

key express

- They say the rifle is the second life of a soldier.

ex) They say the _______ is the second life of 이승엽.

They say the _______ is the second life of cook.

They say the _______ is the second life of _______.

- keep it in good shape. in good shape VS out of shape

- semiautomatic shoulder weapon :

- ammunition/ammo :

Relay Quiz

방독면?

Military English 7-I will finish it by zero nine hundred hours, tomorrow.

13 Today's Dialogue I

A : Have you finished what I asked you before?

B : Almost done, sir. I will finish it by five thirty.

A : Sorry to ask you but can you do this for me, too?

B : You don't have to hurry up.
You can finish it by ten hundred hours, tomorrow.

A : Ten hundred hours???

14 Today's Dialogue II

A : You don't have to hurry up.
You may finish it by ten hundred hours, tomorrow.

B : Don't worry, sir. I will take care of it.

A : But the sooner the better.

B : I will finish it by zero hundred hours, tomorrow.

A : Thank you, Anyway. Take your time.

B : You know me, sir. It's a piece of cake.

Today's title

key express

- I will finish it by zero nine hundred hours, tomorrow.

- ten hundred hours :

I will finish it by zero ____________ hours, tomorrow. (08 : 20)

I will finish it by zero ____________ hours, tomorrow. (15 : 00)

	일반영어	군사영어
정오		
자정		

- Daily Routine

0600(zero six hundred hours) : reveille

0630(zero six thirty hours) : morning roll call

0700(zero seven hundred hours) : mess call

21 : 30(twenty one thirty hours) : evening roll call

Relay Quiz

외출?

Military English 8-I want to speak with Captain John.

15 Today's Dialogue I

A : This is Private Suh.

B : This is Captain Brown, Bravo Company.

A : Good morning, Captain Brown, May I help you?

B : May I speak to Captain John, Alpha Company?

A : Pardon me, Sir?

B : I want to speak with Captain John.

A : Juliet–Oscar–Hotel–November! You got it?

B : Juliet–Oscar….
Sorry but there is no one by that name in my Company, Sir.

16 Today's Dialogue II

A : Pardon me, Sir?

B : I want to speak with Captain John.
Juliet–Oscar–Hotel–November! You got it?

A : Ah, Captain John, my Company Commander!

B : Yes, that's it, It's urgent. Connect him to me right away.

A : Yes, Sir. Wait just a second.

Today's title

key express

- I want to speak with Captain John.

  = May(Can) I speak with Captain John?

- 전화대화

  May I have your name, Please?

  Would you hold the line, Please?

  ○○○ is on another line at the moment.

  Okay, I call her back later then, Bye.

Relay Quiz

군번줄?

Military English 9-I don't have any weapons to harm you with, either.

17 Today's Dialogue I

A : Why did they start saluting this way?

B : Well, I guess it started from the way people took off their caps, like this? Then do you know how it started?

A : I don't know, either.
Anyhow why did they greet this way?
That's quite interesting to me.

Hat trick?

18 Today's Dialogue II

A : I don't have any weapons.
And I don't have any intention to harm you with, either.

B : Same here. I don't have any weapons to harm you with, either.

A : However you know, every part of my body is a lethal weapon.

Today's title

key express

- I don't have any weapons to harm you with, either.

- Do you speak ____ foreign languages?
- Did you make ____ New Year's resolutions?
- He did better tha ____ before him.
- However you know, every part of my body is a lethal weapon.
- Every part of my body is healthy.

Relay Quiz

(웃)주름?

Military English 10-Anyway, I had a chance to be on board an aircraft carrier.

19 Today's Dialogue I

A : Long time no see, Sergeant Bryan.
How have you been?

B : Long time no see, Private Suh.
I have been to a naval base.

A : Naval base! You must've been sightseeing around the sea.
I envy you.

B : I went there for work not for sightseeing.
Anyway, I had a chance to be on board an aircraft carrier.

A : Aircraft? carrier? Is it something like Starcraft?

B : Not a game!

craft : aircraft spacecraft starcraft aircraft carrier

20 Today's Dialogue II

A : I went there for work not for sightseeing.
Anyway, I had a chance to be on board an aircraft carrier.

B : I wish I could be on board a carrier, someday.

A : Why don't you go there for me next year?

B : That's a good idea.

Today's title

key express

- Anyhow, I had a chance to be on board an aircraft carrier.

- ______________ have a lunch with the battalion commander a few days ago.
- ______________ hear a famous jazz singer's songs at a bar yesterday.

- _______ I could be on board a carrier, someday.
- _______ everyday could be like Christmas.
- _______ everyday could be like a leave.

Relay Quiz

암구호?

Military English 11-Would you give me an overnight pass considering my service to my platoon?

21 Today's Dialogue I

A : Private Suh, how have you been doing these days?

B : Everything's OK, sir.
My seniors are also good to me.

A : It is good that you always look so happy.

B : Thank you, sir. you know, I'm always cheering up my platoon.

A : Private, Suh. Will you move to my left side?

B : Yes, sir. But is there something wrong?

22 Today's Dialogue II

A : It is good that you always look so happy.

B : Thank you, sir. you know, I'm always cheering up my platoon.
Would you give me an overnight pass considering my service to my platoon?

A : I will think about it.

B : May I take it as your promise, sir?

A : Yes, you may, I promise.

Today's title

key express

- Would you give me an overnight pass considering my service to my platoon?
- He is the life of the party.
- He knows how to party.
- Let's liven things up.
- He is a wet blanket.

Relay Quiz

군견?

Military English 12-I request fire support, over.

23 Today's Dialogue I

A : Black Hawk Zero, Black Hawk Zero. Fire request, over.

B : Black Hawk One, Black Hawk One. Say again, over.

A : Enemy troops are approaching. Friendly forces are in great danger.
I request urgent fire, over.

B : Black Hawk One, Black Hawk One. I can't hear you. Say again, over.

A : I request fire support, over.

B : Radio check. Fires will be on Hill 326 in five minutes. Radio Check.

A : Radio Check? What are you talking about? Where is the radio?

Radio check :

24 Today's Dialogue II

A : I request fire support, over.

B : Radio check. Fires will be on Hill 326 in five minutes. Radio Check.

A : Roger.

B : Take cover from the friendly fire immediately. Read back.

A : I read back. Fires will be on Hill 326 in five minutes.
Take cover from the friendly fire immediately, over.

B : Your company will be relieved soon. Stand firm for a while, over.

A : I have strong trust in friendly fire. out.

Today's title

key express

- I request fire support, over.

A : I'm lonely. I request you set me up with a pretty woman. over.

B : Radio check. The girls I know don't speak Korean.
They don't even understand Korean.
I recommend you improve your English. Radio check.

A : Roger.

B : Enjoy English and practice your English! Read back.

A : I read back. Enjoy English and practice your English!

B : One more thing. Watch Military English, out.

Relay Quiz

PT체조?

Military English 13-He can see nothing but the sea when he goes far away.

25 Today's Dialogue I

A : Tell me about your temporary duty on the naval base. please.

B : Didn't I say that I was on board an aircraft carrier?

A : How wonderful to be in the Navy.
They can travel around the world on a nice ship.

B : A friend of mine who is working for the navigation branch said navigation branch he would miss the land so much after working so many months on the sea.

A : Navigation?

26 Today's Dialogue II

A : A friend of mine who is working for the navigation branch said navigation branch he would miss the land so much after working so many months on the sea.

B : I can understand him. He can see nothing but the sea when he goes for away.

A : And to tell the truth, I …. I was seasick when I was on board an aircraft carrier. To tell the truth, I would never be on board a ship, any ship.

B : Please don't sit next to me on the bus.

Today's title

key express

- He can see nothing but the sea when he goes far away.

- He did __________ complain when he was in an army training camp.
- I do __________ watch TV on weekends.

- 뱃멀미 :
- 차멀미 :

- overeat/throw up
- I feel like throwing up.

Relay Quiz

야전상의?

Military English

실용영어
(Practical English)

be동사와 인칭대명사

› **UNIT 1.** be동사와 인칭대명사

› **UNIT 2.** be동사의 부정문과 의문문

› **UNIT 3.** 인칭대명사의 격변화

UNIT 1 be동사와 인칭대명사

Prestudy

▣ 다음을 해석하시오.

1. They are classmates.

2. She is 14 years old.

3. I'm smart.

4. They were in the park.

▣ 다음을 영작하시오.

1. 나는 가수이다.

2. 그는 잘생겼다.

3. 우리는 어제 부산에 있었다.

Key Point

1. **be동사 :** ~이다. ~있다.

2. **인칭대명사 :** 사람/사물을 지칭하는 말

<table>
<tr><th>인칭</th><th>단수</th><th>be동사</th><th>줄임말</th><th>복수</th><th>be동사</th><th>줄임말</th></tr>
<tr><td>1인칭</td><td>I (나는)</td><td>am
was</td><td>I'm</td><td>we (우리)</td><td rowspan="7">are
were</td><td rowspan="7">We're
you're
they're</td></tr>
<tr><td>2인칭</td><td>You (너는, 당신은)</td><td>are
were</td><td>you're</td><td>You (당신들, 너희들)</td></tr>
<tr><td rowspan="5">3인칭</td><td>he (그는)</td><td rowspan="5">is
were</td><td rowspan="5">he's
she's
it's
that's
* this is는 줄여 쓰지 않는다.</td><td rowspan="2">they (그들은, 그것들은)</td></tr>
<tr><td>she (그녀는)</td></tr>
<tr><td>it (그것은)</td><td rowspan="2">these (이것들은)</td></tr>
<tr><td>this (이것은)</td></tr>
<tr><td>that (저것은)</td><td>those (저것들은)</td></tr>
</table>

1) I am~. [나는 ~이다.]

I am a student. [나는 학생이다.]=I'm a student.

줄임말은 be동사의 앞 글자를 없애고 그 자리에 [']를 붙여준다.

2) You are~. [너는(너희들은) ~이다.]

You are Kate. [너는 케이트이다.]=You're Kate.

You are students. [너희들은 학생들이다.]=You're students.

3) He is~. [그는 ~이다.]/She is~. [그녀는 ~이다.]/It is~. [그것은 ~이다.]

3인칭 단수 : 제 3자나 사물을 말한다. (he, she, this, that, it, Jane, Kate…)

He is my father. [그는 나의 아버지시다.]=He's my father.

She is a teacher. [그녀는 선생님이다.]=She's a teacher.

It is my book. [그것은 나의 책이다.]=It's my book.

4) This is ~. [이것은/ 이 사람은 -이다.]/That is ~.[저것은/ 저 사람은 -이다.]

① This는 가까이에 있는 사람이나 사물을 가리킬 때 쓰인다.

This is a clock. [이것은 시계이다.]

This is my sister. [이 사람은 나의 여동생이다.]

* this와 is는 줄여 쓰지 않는다.

② That은 멀리 있는 사람이나 사물을 가리킬 때 쓰인다.

That is a car. [저것은 자동차이다.]=That's a car.

That is my mother. [저 사람은 나의 어머니시다.]=That's my mother.

5) We are ~. [우리는 -이다.]/They are ~. [그들은 -이다.]

We are doctors. [우리는 의사들이다.]=We're doctors.

They are teachers. [그들은 선생님들이다.]=They're teachers.

They are my dogs. [그것들은 나의 개들이다.]=They're my dogs.

※ 〈There is+단수〉 There is a computer.

〈There are+복수〉 There are many cars on the road.

Basic Test

※ ( ) 안에서 알맞은 말을 고르시오.

01 There (is, are) a box in the room.

02 I know you. You (am, are) in my class.

03 The girls (was, were) in the park at that time.

04 I (was, were) 160cm tall last year.

※ 다음 〈보기〉에서 알맞은 말을 골라 쓰시오.

보기	am, are, is, was, were

05 I go to school. I ______ a student.

06 I have a cup. It ______ on my desk.

07 I like my classmates. They ______ kind to me.

08 He's at home tonight, but he ______ at the concert last night.

09 You're in class today, but you ______ at the party yesterday.

※ 우리말과 뜻이 같도록 빈칸에 알맞은 말을 쓰시오.

10 형은 26살이다. 그는 미혼이다.

My brother ________ 26 years old.
________ ________ single.

11 나는 책을 5권 가지고 있다. 그것들은 책상 위에 있다.

I have five books. _____ _____ on my desk.

12 Susan과 나는 한 아파트에서 산다. 우리는 룸메이트이다.

Susan and I live in an apartment.
________ ________ roommates.

13 TV에서 마술쇼가 있었다. 그것은 매우 재미있었다.

There _____ a magic show on TV.
_____ _____ very interesting.

Practice Test

※ 다음 빈칸에 들어갈 알맞은 말을 고르시오.

01 My father was a pilot. Now I _______ a pilot, too.

① is ② are
③ am ④ were
⑤ was

02 There _______ a new game CD on my desk yesterday.

① is ② are
③ am ④ were
⑤ was

03 Jill and I _________ at the cinema last Saturday.

① is ② are
③ am ④ were
⑤ was

04 다음 빈칸에 들어갈 수 없는 말은?

_______ were at Rain's concert yesterday.

① We ② She
③ They ④ Mr. and Mrs. Smith
⑤ Hyolee and Boa

05 다음 밑줄 친 부분 중 어법상 틀린 것은?

① Ron is my close friend.
② They were at the restaurant now.
③ There are many balls in the basket.
④ Mimi and Sora are in the classroom.
⑤ She's with my English teacher.

06 다음 중 어법상 맞는 문장은?

① She's angry yesterday.
② They're all handsome.
③ Janet were very pretty.
④ I were with Sarah last evening.
⑤ They am at the restaurant.

07 다음 빈칸에 들어갈 말이 순서대로 바르게 짝지어진 것을 고르시오.

• Karen and you ______ late for school again yesterday.
• Seungyeop ______ a great baseball player.

① were–were ② were–are
③ were–is ④ are–is
⑤ are–were

Reading Comprehension

01 다음 글의 내용과 일치 하지 <u>않는</u> 것은?

> At a baseball game, Bob was surprised to see that a man had two seats. He was using one just to keep his things on. After a few minutes, Bob said to the man, "Did you buy two tickets just for yourself?" "No, one is my wife's seat," the man answered, "but she died."
> "Oh, I'm terribly sorry to hear that," Bob said. "But couldn't you give the seat to one of your relatives?" "Well, I tried," the man answered. "But they're all at the funeral."

① 야구 경기장에서 일어난 일이다.
② 남자가 두 개의 좌석을 차지하고 있었다.
③ 남자의 아내는 화장실에 갔다.
④ 남자는 한 좌석에 물건을 올려 놓았다.
⑤ 남자의 친척들은 장례식에 갔다.

02 다음 글에서 밑줄 친 <u>these kinds of games</u>가 뜻하는 것은?

> Playing computer games helps kids to become familiar with computers. Most computer games are not violent. But some games let you shoot a gun at people on the screen. When someone is hit, blood and body parts fly into the air. Another game shows a monster eating a person. Some children like to fight and they use the computer to pretend to fight. Parents are worried that <u>these kinds of games</u> are not good for children to play. They want less violence in the computer games.

① 학습용 컴퓨터 게임
② 성인용 컴퓨터 게임
③ 소녀 취향의 컴퓨터 게임
④ 폭력적이고 잔인한 컴퓨터 게임
⑤ 어린이들에게 인기가 많은 게임

UNIT 2 be동사의 부정문과 의문문

Prestudy

▣ 다음을 해석하시오.

1. I'm not bored.

____

2. A : Is he your father?

____

B : No, he isn't.

____

▣ 다음을 영작하시오.

1. 그는 배고프지 않다.

____

2. A : 그들은 야구 선수들이니?

____

B : 응, 그래.

____

Key Point

1. be동사의 부정문 : [주어+be+not] ~이 아니다. ~없다.

주어	be동사+not	줄임말 1	줄임말 2
I	am not	I'm not	–
YOU	are not	You're not	You aren't
HE/SHE/IT	is not	He's/She's/It's not	He/She/It isn't
WE/YOU/THEY	are not	We're/You're/They're not	We/You/They aren't

I am not Mike. [나는 Mike가 아니다.]
You are not a teacher. [너는 선생님이 아니다.]=You aren't a teacher.
He is not tall. [그는 키가 크지 않다]=He isn't tall.
She is not a teacher. [그녀는 선생님이 아니다.]=She isn't a teacher.
We are not friends. [우리는 친구들이 아니다.]=We aren't friends.

2. be동사의 의문문 : [be동사+주어…?] ~ 입니까? ~있습니까?

be동사+주어…?	긍정대답	부정대답
Am I …?	Yes, you are.	No, you aren't.
Are you …?	Yes, I am.	No, I'm not.
Is he/she/it …?	Yes, he/she/it is.	No, He/she/it isn't.
Are we/they/you …?	Yes, you/we/they are.	No, you/we/they aren't.

You are a doctor. [당신은 의사이다.]
→Are you a doctor? [당신은 의사입니까?] : be동사인 are만 문장 앞으로 보내준다.
Yes, I am./No, I am not. : Are you ~?로 물으면 I am으로 답한다.
He is a teacher. [그는 선생님이다.]
→Is he a teacher? [그는 선생님입니까?] : be동사인 is만 문장 앞으로 보냄

Yes, he is./No, he isn't.

She is beautiful. [그녀는 아름답다.]

→Is she beautiful? [그녀는 아름답습니까?] : be동사인 is만 문장 앞으로 보냄

Yes, she is./No, she isn't.

They are your friends. [그들은 당신의 친구들이다.]

→Are they your friends? [그들은 당신의 친구들입니까?] : be동사인 are만 문장 앞으로

※ 의문문에 he, she, they가 있으면 그대로 he, she, they로 답한다.

Basic Test

※ (   ) 안에서 알맞은 말을 고르시오.

01 (Is, Are) John and Mary late?

02 They (are not, not are) good friends.

03 (Was, Were) he your middle school teacher?

04 Nick (is not, not is) a computer programmer.

※ 다음 밑줄 친 부분을 바르게 고치시오.

05 Is these your pens?

06 Mary and John isn't British. They're American.

07 Jill is busy yesterday.

08 You're in class today, but you aren't in class yesterday.

※ 다음 주어진 문장을 지시에 따라 바꾸시오.

09 Minhee was at the dance party.

→(부정문) ______________________

→(의문문) ______________________

10 Mike and Susie are from England.

→(부정문) ______________________

→(의문문) ______________________

※ 우리말과 뜻이 같도록 주어진 단어를 배열하시오.

11 그 사람들은 중국 출신이 아니다.
(not, are, from, China, they)

______________________________.

12 지난 금요일에 그는 사무실에 있었습니까?
(last Friday, was, in, he, his office)

______________________________?

13 그녀가 당신의 누나입니까?
(your, she, sister, is)

______________________________?

New Words British 영국인의

Practice Test

01 다음 질문에 대한 알맞은 응답을 고르시오.

A : Is it your computer?
B : ______________________

① Yes, it is. ② No, it is.
③ Yes, you are. ④ No, you aren't.
⑤ Yes, it isn't.

02 다음 질문에 대한 알맞은 응답을 고르시오.

A : Are you really Tony's sister?
B : ______________________

① Yes, I am. ② No, you aren't.
③ Yes, I'm not. ④ Yes, you are.
⑤ No, we are.

03 다음 질문에 대한 알맞은 응답을 고르시오.

A : Are your father and mother in the room?
B : ______________________

① Yes, we are. ② No, they isn't.
③ No, they aren't. ④ No, we aren't.
⑤ Yes, they aren't.

04 다음 빈칸에 들어갈 수 없는 말은?

Were ______ with Samsoon at the dance party last night?

① you ② they
③ he ④ Tim and Brian
⑤ her parents

05 다음 우리말을 영어로 바르게 옮긴 것을 고르시오.

그녀는 지난 토요일에 어디에 있었나요?

① Where she were last Saturday?
② Where is she last Saturday?
③ Where she is last Saturday?
④ Where was she last Saturday?
⑤ Where were she last Saturday?

06 다음 우리말을 영어로 바르게 옮긴 것을 고르시오.

용준이는 어젯밤 집에 없었다.

① Yongjun is at home last night.
② Yongjun isn't at home last night.
③ Yongjun was at home last night.
④ Yongjun wasn't at home last night.
⑤ Yongjun weren't at home last night.

Reading Comprehension

01 다음 글의 내용으로 보아 'I'에 대한 부모님의 태도는?

My mom and dad won't let me do any of the things my friends do. For example, on Saturday evenings everyone goes to the disco. But my parents say I have to be home by nine o'clock. It's really unfair. Everyone laughs at me and calls me a mummy's boy. Also my mom goes into my bedroom and looks through my things when I'm at school. When I ask her not to, she says that I'm too young to have a private life. They treat me as a baby. And they don't trust me to do things on my own.

① 늘 야단친다.
② 관심이 없다.
③ 간섭이 심하다.
④ 친구처럼 대한다.
⑤ 사랑이 부족하다.

02 다음 글의 밑줄 친 (A)와 (B)의 관계로 알맞은 것은?

(A)<u>I</u>'m writing to (B)<u>you</u> about your son, Paul. His school work has been very poor recently. Last week his answer to a question on the history test was that Australia was discovered by Columbus. And on a geography test, he answered that coffee comes from supermarkets! I don't know what is wrong with him. But I'm sure he isn't studying hard. I'm worried about him. So how about coming in to talk to me about it? Please come to see me as soon as possible.

	(A)	(B)
①	학부모	학생
②	학부모	선생님
③	선생님	학부모
④	선생님	학생
⑤	학생	선생님

UNIT 3 인칭대명사의 격변화

Prestudy

▣ 다음을 해석하시오.

1. You and I are just friends.

2. His legs are long.

3. He likes me.

4. The diamond ring isn't mine.

▣ 다음을 영작하시오.

1. 그는 영리하다.

2. 그녀의 미소는 아름답다.

3. 나의 부모님은 나를 사랑하신다.

4. 책상 위에 있는 편지는 너의 것이다.

Key Point

1. 인칭대명사의 격변화

사람이나 사물의 이름을 대신하는 인칭대명사는 그 역할(격)에 따라 4가지 형태가 있다.

		주격	소유격	목적격	소유대명사
		~는, ~가	~의	~를, ~에게	~의 것
1인칭	단수	I	my	me	mine
	복수	we	our	us	ours
2인칭	단수	you	your	you	yours
	복수	you	your	you	yours
3인칭	단수	he	his	him	his
		she	her	her	hers
		it	its	it	–
	복수	they	their	them	theirs

2. 주격, 소유격, 목적격, 소유대명사의 쓰임

주격+동사	She and I are friends. →We are friends. You and she are pretty. →You are pretty. You and I are tall. →We are kind. Tom and Mary are kind. →They are kind.
소유격+명사	This is my dog. I like his sister.
동사/전치사+목적격	I like Mary. →I like her. He gives the book to Mike. →He gives the book to him.
소유격+명사→소유 대명사	That is my book. →That book is mine. The cat is our cat. →The cat is ours.

3. 대명사의 격은 문장에서 알맞게 사용한다.

① 주격 : 문장에서 주어 역할 [~은, 는, 이, 가]

② 소유격 : 소유를 나타내는 명사 앞 [~의]

③ 목적격 : 동사 또는 전치사의 목적어 역할 [~을, ~를]

④ 소유대명사 : 소유격+명사를 대신하는 역할 [~의 것]

Basic Test

※ ( ) 안에서 알맞은 말을 고르시오.

01 (You, Your) are my best friends.

02 Happy is their dog. It's (them, theirs).

03 These are her books. They are (her, hers).

04 He loves (we, us) a lot. We love (he, him), too.

※ 다음 <보기>에서 알맞은 말을 골라 쓰시오.

보기	your, she, hers, him, mine

05 This is my car. It's ______________.

06 That's his bag, not ____________ bag.

07 I know __________. He's from Canada.

08 _____ likes him very much. He's handsome.

※ 우리말과 뜻이 같도록 빈칸에 알맞은 말을 쓰시오.

09 지영이는 내 친구이다. 나는 그녀를 좋아한다.

Jiyoung is _____ friend. I like _________.

10 그것들은 너의 신발이다. 그것들은 너의 것이다.

They're _____ shoes. They're _________.

11 그녀는 우리들의 영어 선생님입니다. 우리들은 그녀를 매우 좋아합니다.

She is _____ English teacher. _______ like very much.

※ 우리말과 뜻이 같도록 주어진 단어를 배열하시오.

12 너의 자전거는 새 것이구나. 내 것은 오래된 것이다. (old, mine, is)

Your bike is new. ____________________

__________________.

13 Neo는 Trinity를 사랑한다. 그녀도 그를 사랑한다. (she, him, loves, too)

Neo loves Trinity. ____________________

__________________.

Practice Test

01 다음 빈칸에 들어갈 알맞은 말을 고르시오.

> She's ________ elder sister.

① him ② hers ③ his
④ he's ⑤ yours

02 다음 빈칸에 들어갈 알맞은 말을 고르시오.

> Jina is a pretty girl. I love ______________.

① she ② her ③ hers
④ Jina's ⑤ them

03 다음 빈칸에 들어갈 알맞은 말을 고르시오.

> A : Is this cell phone yours?
> B : Yes, it is ________.

① him ② his ③ my
④ mine ⑤ hers

04 다음 우리말을 영어로 바르게 옮긴 것을 고르시오.

> 나는 그녀의 친구들을 알고 있다.

① I know her friends.
② I know friends her.
③ I know she friends.
④ I know friends she.
⑤ I know hers friends.

05 다음 우리말을 영어로 바르게 옮긴 것을 고르시오.

> 그 전화기들과 가방들은 그들의 것이다.

① Their the phones and bags are them.
② They're the phones and bags.
③ The phones and bags are them.
④ The phones and bags are theirs.
⑤ The phones and bags are they're.

06 다음 밑줄 친 부분 중 어법상 <u>틀린</u> 것을 고르시오.

① Brenda is <u>her</u> sister.
② Mary loves <u>hers</u> small dolls.
③ All his books are <u>mine</u> now.
④ I like <u>their</u> songs.
⑤ That pretty girl is <u>our</u> daughter.

07 다음 밑줄 친 부분 중 어법상 <u>틀린</u> 것을 고르시오.

① <u>Their</u> new songs are really cool.
② This phone is not <u>her</u>. It's mine.
③ This MP3 player is cute. Is it <u>yours</u>?
④ Is she <u>your</u> sister? She's beautiful.
⑤ She is <u>Tim's</u> girlfriend.

Reading Comprehension

01 다음 글의 제목으로 가장 알맞은 것은?

In 1941, there was a big war in Europe and many soldiers were fighting. Mrs. Vial's son was also in the army. She decided to help by baking hundreds of cakes to send to the soldiers. When she finished baking, Mrs. Vial lost her wedding ring. It must have fallen into one of cakes! But she didn't want to cut open the cakes, so she wrote a note to go with each one. The note asked everyone who ate a cake to look for her ring inside. Then one day a soldier returned it. Who was the soldier who found the ring? It was her son, Sgt. Ronnie Vial!

① A famous Baker, Mrs. Vial.
② Origins of the Ring
③ How a Lost Ring was Found
④ How to Bake a Cake
⑤ The Meaning of the Wedding Ring

02 다음 글의 제목으로 알맞은 것은?

There are many fighting sports enjoyed all over the world. One of the simplest of these was practiced in England until the seventeenth century. It was shin-kicking. Players fought each other wearing boots with iron toes. The rules were quite simple. Each player kicked at the shins of the other. The game ended when one couldn't stand or gave up. After fighting, players had trouble walking.

① What Shin-kicking Is
② Why Shin-kicking Is Cruel
③ Why Shin-kicking Was Popular
④ How Many Fighting Sports There Are
⑤ How the Rules of Shin-kicking Were Made

New Words Sgt. 하사관, 병장
shin-kicking 정강이 차기

일반동사

UNIT 1 일반동사의 의미와 형태

Prestudy

▣ 다음을 해석하시오.

1. I drink milk every morning.

___

2. You look nice.

___

3. My mother works for a bank.

___

▣ 다음을 영작하시오.

1. 나는 너의 도움이 필요해.

___

2. 너는 영어를 잘 말하는구나.

___

3. 그녀는 음악을 좋아한다.

___

Key Point

1. **일반동사 :** 일반동사는 be동사와 조동사를 제외한 나머지 동사로 동작이나 상태를 나타내는 동사이다.

be동사	am, are, is, was, were
조동사	will, may, can, must, should
일반동사	eat, run, play 등 have, know, like 등

1) 동사는 주어 다음에 와서 [주어+동사+목적어]의 어순이 된다.
I have a guitar. [나는 기타를 가지고 있다.]
I speak English. [나는 영어를 말한다.]
You play basketball. [너는 농구경기를 한다.]

2) 주어가 3인칭 단수인 경우 : 일반 동사의 경우에는 동사 뒤에 -(e)s를 붙인다.
ex) Jane, Minsu, Mike, a boy(소년), a girl(소녀) 등 사람의 이름이나 사람
ex) a dog(개), a book(책), a desk(책상) 등 동물이나 사물 하나
* 1인칭인 I[나는]와 2인칭 You(너는)를 제외한 he(그는), she(그녀는), it(그것은), this(이것은), that(저것은) 등 인칭대명사와 제3의 사람 한 명이나 물건 하나를 3인칭 단수라고 한다.

2. **3인칭 단수형 동사 만들기**

일반적으로 대부분의 일반 동사의 끝에 -s를 붙인다.	eat→eats, want→wants He speaks English. She sings Christmas songs.
동사가 -o, -s, -ch, -sh로 끝나면 -es를 붙인다.	go→goes, watch→watches, wash→washes I go to school by bus. Mike goes to school by bus. She goes to school by bus.
동사가 [자음+y]로 끝나면 y를 i로 고치고 -es를 붙인다. * [모음+y]이면 s만 붙임 play→plays	[자음 : b, c, d, f, g, h, j, k, l, m, n, ……] study→studies, cry→cries, try→tries He studies English.
have는 has로 바뀐다.	He has a ball. Jane has a beautiful house. * 주어 He(그는)가 3인칭 단수이므로 have를 쓰면 안 되고 has를 써야함

Basic Test

※ (　) 안의 단어를 알맞은 형태로 고치시오. (단, 동사의 현재형으로 쓰시오.)

01 Jina (have) a good digital camera.

02 My uncle is a scientist. He (make) the space shuttle.

03 Jongsu is good at English. He (speak) with foreigners in English.

※ 다음 〈보기〉에서 알맞은 말을 골라 적절히 바꾸어 쓰시오. (단, 동사의 현재형으로 쓰시오.)

보기	do, go, kiss, fix

04 He ______ his homework after supper.

05 Heejin ______ to the supermarket and buys some milk.

06 His daughter ______ him every morning.

07 My father ______ cars very well.

※ 우리말과 뜻이 같도록 빈칸에 알맞은 말을 쓰시오.

08 그 아기는 매우 자주 운다.

The baby ________ very often.

09 진호는 매주 일요일에 엄마를 돕는다.

Jinho _____ his mom every Sunday.

10 아버지는 매일 아침마다 산책을 하신다.

My father _____ a walk every morning.

11 내 친구는 피아노를 아주 잘 친다.

My friend _____ the piano very well.

※ 밑줄 친 부분을 (　) 안의 말로 바꾸어 문장을 다시 쓰시오.

12 <u>We</u> drink orange juice every morning. (He)

→ He ______________________.

13 <u>You</u> have a lot of work. (She)

→ She ______________________.

New Words space shuttle 우주왕복선
foreigner 외국인

Practice Test

01 다음 중 동사의 3인칭 단수 현재형이 잘못 연결된 것은?

① make makes　　② go−goes
③ watch−watches　　④ study−studys
⑤ have−has

※ 다음 빈칸에 들어갈 알맞은 말을 고르시오.

02 ______ studies English every evening.

① They　　② I　　③ She
④ We　　⑤ You

03 ______ speak Korean very well.

① Amy　　② She　　③ He
④ Brian　　⑤ They

04 다음 우리말을 영어로 바르게 옮긴 것은?

> 다나는 매일 피아노를 친다.

① Dana play the piano every day.
② Dana playes every day the piano.
③ Dana plays the piano every day.
④ Dana playes the piano every day.
⑤ Dana play every day the piano.

05 다음 밑줄 친 부분 중 어법상 틀린 것을 고르시오.

① Tony loves Jane so much.
② My sister enjoys computer games.
③ Her brothers often fights.
④ Jenny lives in Busan now.
⑤ Amy goes to church every Sunday morning.

06 다음 밑줄 친 부분 중 어법상 틀린 것을 고르시오.

① Tony has two brothers and a sister.
② I help my mother all the time.
③ She hates hamburgers and pizza.
④ You needs some help.
⑤ Charlie does two things at the same time.

07 다음 중 빈칸에 공통으로 들어갈 알맞은 말은?

> • He ______ Harry Potter DVDs at home.
> • Harry ______ dinner at 7o'clock every day.

① has　　② have　　③ eat
④ eats　　⑤ watches

Reading Comprehension

01 다음 글의 제목으로 알맞은 것은?

It was one hundred years ago in New York. A street seller named Harry Stevens sold "dachshund sausages" in rolls. A dachshund is a German dog and he named the sausages after this long, thin dog. In 1901, a cartoonist drew a cartoon showing one of Mr. Stevens' sausages. But he didn't know the spelling of "dachshund," so he just called it a "hot dog." The name became famous and people have used it until now.

① What's a "Hot Dog" like?
② What Does "Dachshund" Mean?
③ How Did the "Hot Dog" Get Its Name?
④ Why Did the "Hot Dog" Become Famous?
⑤ How Did the Sausage Come to the US?

02 다음 글의 He가 밑줄 친 부분과 같이 행동한 이유로 알맞은 것은?

A friend of mine had a bad day at work. Being upset, he didn't speak much as I drove him home. Then as we walked toward his house, I saw him <u>touching a small tree both hands</u>. A moment later, his face seemed bright and happy. Entering his house, he hugged his two small children, and kissed his wife. The next day, I asked him about the tree. "Oh, that's my trouble tree," he answered. "I often have troubles at work, but the troubles don't belong to my wife and children. So I just hang them up on the tree every night when I come home. Then leaving home, I pick them up again. And interestingly, there are less troubles than the day before."

① 나무가 얼마나 컸는지 알아보려고
② 직장에서의 걱정거리를 달아 놓으려고
③ 나무가 자기 아내인 줄로 잘못 알아서
④ 나무를 사랑과 정성으로 키우려고
⑤ 나무에게 주문을 걸기 위해서

UNIT 2 일반동사의 과거형

Prestudy

▣ 다음을 해석하시오.

1. She played golf yesterday.

____

2. I saw her in the bookstore.

____

3. Jane invited me to her birthday party.

____

▣ 다음을 영작하시오.

1. 나는 어젯밤에 영화를 봤다.

____

2. 나는 유럽 여행을 마치고 집으로 돌아왔다.

____

Key Point

1. **일반동사의 과거형 만드는 법 :** 동사는 그 모양을 변화시켜 시간을 나타낼 수 있다. 일반 동사는 과거형을 사용하여 과거시제를 나타낸다. 동사의 과거형을 만드는 방법에는 규칙변화와 불규칙 변화가 있다.

1) 규칙변화의 형태

동사원형+ed	want→wanted talk→talked	help→helped learn→learned
e로 끝나면+d	like→liked love→loved	live→lived smile→smiled
자음+y로 끝나면 y를 'i'로 바꾸고 +ed	study→studied dry→dried	cry→cried
단모음+단자음으로 끝나고 모음에 강세가 있으면 자음을 한 번 더 쓰고+ed	stop→stopped chat→chatted	drop→dropped step→stepped

2) 불규칙변화

형태가 변하는 것	do→did have→had take→took buy→bought break→broke draw→drew find→found hear→heard send→sent say→said speak→spoke tell→told win→won	go→went eat→ate give→gave bring→brought build→built fall→fell forget→forgot hide→hid lose→lost sell→sold spend→spent throw→threw write→wrote	run→ran come→came know→knew sit→sat catch→caught drink→drank get→got keep→kept mean→meant sing→sang stand→stood wake→woke	see→saw make→made meet→met fly→flew choose→chose feel→felt grow→grew lend→lent ride→rode sleep→slept swim→swam wear→wore
원형과 모양이 같은 것	cut→cut hit→hit let→let put→put			

※ 과거형은 과거 시간을 나타내는 표현과 함께 주로 쓰인다.

yesterday, last night/week/year/month, two hours ago, in 2003 등

Basic Test

※ 우리말과 뜻이 같도록 밑줄 친 단어를 바르게 고치시오.

01 우리는 3년 전에 만났다.
We <u>meet</u> three years ago.

02 그는 어제 은행에 갔다.
He <u>goes</u> to the bank yesterday.

03 그녀는 2시간 전에 떠났다.
She <u>leaves</u> two hours ago.

※ 우리말과 뜻이 같도록 〈보기〉에서 알맞은 말을 골라 적절히 바꾸어 쓰시오.

보기	have, watch, read, go, make

04 나는 지난주에 해리포터를 읽었다.
I _____ Harry Potter last week.

05 Ann은 어제 친구들과 함께 저녁을 먹었다.
Ann ______ dinner with her friends yesterday.

06 현주는 비행기 안에서 영화를 보았다.
Hyeonju _____ a movie in the airplane.

07 나는 지난 주말에 애플파이를 만들었다.
I _____ an apple pie last weekend.

※ ( ) 안에서 알맞은 말을 고르시오.

08 Andrew (came, comes) home late yesterday.

09 I (enjoy, enjoyed) the party last night.

10 Columbus (arrived, arrives) in America in 1492.

※ 밑줄 친 부분에 유의하여 우리말로 옮기시오.

11 Mr. Smith <u>woke</u> up early this morning.

____

12 My father <u>stopped</u> smoking 5 years ago.

____

13 The Korean soccer team <u>played</u> very well in the 2002 World Cup.

____

Practice Test

01 다음 중 동사의 현재형과 과거형이 바르게 연결된 것은?

① stop–stoped ② open–openned
③ try–tried ④ have–haved
⑤ do–doed

※ 다음 빈칸에 들어갈 알맞은 말을 고르시오.

02 Ron ______ a lot of mistakes yesterday.

① make ② made ③ maked
④ makes ⑤ making

03 They __________ a big LCD TV for their parents last Thursday.

① buy ② buys ③ buyed
④ buyied ⑤ bought

04 다음 빈칸에 들어갈 말이 순서대로 바르게 짝지어진 것은?

• Jane _____ to her boyfriend every day.
• John _____ an email to me last friday.

① writes–wrote ② writes–writes
③ wrote–writes ④ writes–write
⑤ write–wrote

※ 다음 우리말을 영어로 바르게 옮긴 것을 고르시오.

05 2003년에 그는 56개의 홈런을 쳤다.

① He hit 56 home runs in 2003.
② He hits 56 home runs in 2003.
③ He hitted 56 home runs in 2003.
④ He is hit 56 home runs in 2003.
⑤ He was hit 56 home runs in 2003.

06 어제 Sharon은 이탈리아 음식을 먹었어.

① Sharon Italian food yesterday ate.
② Sharon eated Italian food yesterday.
③ Sharon Italian food ate yesterday.
④ Sharon ate Italian food yesterday.
⑤ Sharon ated Italian food yesterday.

07 다음 밑줄 친 부분 중 어법상 <u>틀린</u> 것을 고르시오.

① Tony <u>readed</u> the letter last weekend.
② I <u>saw</u> Go Ahrah in front of the KBS building yesterday.
③ Ivy <u>sang</u> her hit songs on stage last night.
④ Brad <u>made</u> friends with Jennifer at the party.
⑤ Angelina <u>met</u> Daniel at her birthday party.

Reading Comprehension

01 다음 글의 내용과 일치하지 <u>않는</u> 것은?

Journalists try hard to take famous people's pictures. However, one journalist tried something different. He looked in Jack Nicholson's trash to get new information. Did he find anything interesting there? Let's see. Mr. Nicholson drinks a lot of beer, a little mineral water, and lots of Seven up. But he doesn't drink any Coke. He doesn't have many meals at home, except for breakfast. He usually has some bacon and perhaps a few peas. When he's hungry between meals, he has some corn chips or a few chocolates.

① 한 기자는 쓰레기를 뒤져 잭 니콜슨에 대한 정보를 얻었다.
② 잭 니콜슨은 맥주와 세븐업을 많이 마신다.
③ 잭 니콜슨은 콜라를 마시지 않는다.
④ 잭 니콜슨은 매끼 식사를 집에서 한다.
⑤ 잭 니콜슨은 간식으로 콘칩과 초코릿을 즐겨 먹는다.

02 다음 글의 주인공은 무엇을 하고 있는가?

It was such a fun experience! We were really high up, and we could see everything. All the houses and people looked really small. It was very quiet as well—but cold! The wind was blowing a little and we moved very slowly. Every few minutes, the pilot turned the fire on to keep us up in the air. There were eight of us in the basket. We were in the air about an hour. I hope to fly in it again.

① 등산
② 묘기 비행
③ 기구 타기
④ 행글라이딩
⑤ 번지 점프

UNIT 3 일반동사의 부정문

Prestudy

▣ 다음을 해석하시오.

1. We don't make noises in the classroom.

____

2. He doesn't study hard.

____

3. Tom didn't come to my birthday party.

____

▣ 다음을 영작하시오.

1. 나는 거짓말을 하지 않는다.

____

2. 그녀는 그를 좋아하지 않는다.

____

3. 그들은 내 이름을 몰랐다.

____

Key Point

1. **일반동사의 부정문** : 현재형은 동사 앞에 don't(doesn't), 과거형은 didn't를 붙인다.

인칭/수	현재/과거 부정형	줄임말
I/You/We/They	do/did not+동사원형	don't/didn't+동사원형
He/She/It 등 3인칭 단수	does/did not+동사원형	doesn't/didn't+동사원형

1) 일반 동사의 부정형 : don't+동사원형 (~하지 않다.)

I have much money. [나는/가지고 있다/많은 돈을]

→I don't have much money. [나는/가지고 있지 않다/많은 돈을]

동사가 have이므로 have 바로 앞에 don't를 붙여주면 된다.

They like classical music. [그들은/좋아한다/고전음악을]

→They don't like classical music. [그들은/좋아하지 않는다/고전음악을]

2) 주어가 3인칭 단수(He, She, It, Mike, The book …)일 경우 : doesn't+동사원형

일반동사 앞에 doesn't를 쓰고 동사에 붙어 있는 s나 es를 없애면 되고 has는 have로 바꿔준다.

She likes soccer. [그녀는/좋아한다/축구를]

→She doesn't likes soccer. [먼저 동사앞에 doesn't를 쓴다]

→She doesn't like soccer. [동사 likes뒤의 s를 없앤다]

[그녀는/좋아하지 않는다/축구를]

He has many friends. [그는/가지고 있다/많은 친구들을]

→He doesn't have many friends. [그는/가지고 있지 않다/많은 친구들을]

3) 일반 동사의 과거형의 부정 : didn't+동사원형

주어에 관계없이 [did not 동사원형]을 사용한다.

He liked soccer very much. [그는/좋아했다/축구를/매우 많이]

→He didn't like soccer very much. [그는/좋아하지 않았다/축구를/매우 많이]

I met her yesterday. [나는/만났다/ 그녀를/어제]

→I didn't meet her yesterday. [나는/만나지 않았다/ 그녀를/어제]

Basic Test

※ ( ) 안에서 알맞은 말을 고르시오.

01 The bank (do not, does not) open on Sundays.

02 He (go not, doesn't go) to bed early.

03 We (don't read, doesn't read) the news paper every day.

※ 다음 밑줄 친 부분을 어법에 맞게 고치시오.

04 Jake doesn't likes comic books.

05 They don't play football last Saturday.

06 Amie don't eat Korean food.

※ 우리말과 뜻이 같도록 ( ) 안의 단어를 이용하여 문장을 완성하시오. (단, 조동사의 부정형은 줄임말로 쓰시오.)

07 그는 오늘 버스를 운전하지 않는다. (drive)

He __________ a bus today.

08 그들은 지금 일하지 않는다. (work)

They __________ now.

09 나는 고양이를 좋아하지 않는다. (like)

I __________ cats.

10 그녀는 어제 숙제를 하지 않았다. (do)

She __________ her homework yesterday.

※ 다음 밑줄 친 부분을 바르게 고치시오. (단, 조동사의 부정형은 줄임말로 쓰시오.)

11 The sun don't rise in the west.

12 He was not brush his teeth before he went to bed last night.

13 They aren't watch television very often.

Practice Test

※ 다음 빈칸에 들어갈 알맞은 말을 고르시오.

01 I ______ have much time.

① do not ② does not
③ was not ④ not do
⑤ not does

02 They ______ smoke or drink now.

① isn't ② wasn't
③ don't ④ doesn't
⑤ didn't

03 My father ______ work late on Fridays.

① isn't ② doesn't
③ wasn't ④ don't
⑤ didn't

04 다음 밑줄 친 부분 중 어법상 틀린 문장을 고르시오.

① We didn't play soccer yesterday.
② Eunji doesn't go to the movies last night.
③ I didn't enjoy the comedy show.
④ This bus doesn't go to Shinchon.
⑤ He didn't study any foreign languages.

05 다음 밑줄 친 부분 중 어법상 틀린 문장을 고르시오.

① Our homeroom teacher does many things for us.
② The musical doesn't start until seven.
③ Ron didn't feel good yesterday.
④ Mary doesn't live here any more.
⑤ I didn't swam in that dirty pool.

06 다음 빈칸에 들어갈 말이 순서대로 바르게 짝지어진 것을 고르시오.

• She's a quiet girl. She ______ talk much.
• Jamie woke up late. So she ______ take a walk this morning.

① don't–doesn't ② doesn't–doesn't
③ doesn't–didn't ④ didn't–didn't
⑤ didn't–doesn't

07 다음 빈칸에 들어갈 말이 순서대로 바르게 짝지어진 것을 고르시오.

• She ______ invite me yesterday.
• I was late. My alarm ______ ring again.

① wasn't–didn't ② wasn't–doesn't
③ didn't–weren't ④ didn't–didn't
⑤ didn't–don't

Reading Comprehension

01 다음 글을 통해 알 수 있는 것은?

A man had a bad pain in his stomach, so he went to see a doctor. The doctor told him to put some salt in warm water and drink it one hour before breakfast. A few days later, he went to see the doctor again. "Are you feeling better?" the doctor asked. "Not yet," answered the man. "I'm actually feeling worse." "Did you drink warm salt water an hour before each day?" the doctor asked. "No," replied the man, " I could only drink it about 15 minutes!"

① 남자는 꾀병을 부리고 있다.
② 의사는 남자에게 처방을 잘못해 주었다.
③ 남자는 의사의 처방을 잘못 알아들었다.
④ 남자는 일주일 후에 상태가 더 좋아졌다.
⑤ 남자는 증세 악화가 의사 때문이라고 생각한다.

02 다음 글에서 Anna Moses에 관한 설명 중, 일치 하지 않은 것은?

Anna Moses was 76 years old. People called her Grandma. But she wanted to do something new. So she began to paint. Grandma Moses painted pictures of houses and trees. She also painted people working and playing. All her pictures were full of bright colors. So people like them. Now her pictures can be seen all over the world. Maybe you've already seen some of them at the gallery.

① 76세의 할머니였다.
② 젊을 때부터 화가로 활동했다.
③ 밝은 색을 이용한 그림을 그렸다.
④ 세계적으로 그녀의 그림이 알려졌다.
⑤ 사람, 집, 그리고 나무 들을 그렸다.

UNIT 4 일반동사의 부정문

Prestudy

▣ 다음을 해석하시오.

1. Do you want milk?—No, I don't.

____

2. Does she get up early in the morning?—Yes, she does.

____

3. Did he like the movie?—Yes, he did.

____

▣ 다음을 영작하시오.

1. 너는 스포츠를 좋아하니?—응, 좋아해.

____

2. 그는 안경을 끼니?—아니, 끼지 않아.

____

3. 그녀는 감기에 걸렸니?—응, 감기에 걸렸어.

____

Key Point

1. 일반동사의 의문문 : Do(es)/Did 주어+동사원형~ ?

주어의 종류	형태	example
1 · 2인칭, 복수	Do+주어+동사원형	I like apples. →Do you like apples?
3인칭 단수	Does+주어+동사원형	She likes oranges. →Does she like oranges?
과거시제	Did+주어+동사원형	He did his homework. →Did he do his homework?

① 주어가 1인칭(I, we), 2인칭(you), 3인칭복수(they, three books, many dogs…)일 경우 문장 앞에 Do만 붙여주면 된다. [Do+주어+동사원형 –?]
You like fish. [너는 생선을 좋아한다.] →Do you like fish? [너는 물고기를 좋아하니?]
–Yes, I do. [그래, 좋아해.]/No, I don't. [아니, 좋아하지 않아]
you(너는)로 질문하면 대답할 때 I(나는)로 받는다.
Do you have many books? [너는 많은 책을 가지고 있니?]
–Yes, I do./No, I don't.

② 주어가 3인칭 단수(he, she, it, Mike, Minho…)일 경우, 문장 앞에 Does를 써주고 뒤에는 동사원형을 쓴다. has의 경우엔 have로 바꿔준다. [Does+주어+동사원형~ ?]
She likes fish. [그녀는/좋아한다/물고기를]
→Does she like fish?–Yes, she does./No, she doesn't.
문장 앞에 Does를 쓰고 likes를 like로 바꿔주면 된다.
Mike exercises every day. [마이크는/운동을 한다/매일]
→Does Mike exercise ever day?–Yes, he does./No, he doesn't.

③ 과거형에서는 Did+주어+동사원형~ ?을 사용한다.
She liked fish. [그녀는 물고기를 좋아했다.]
→Did she like fish?–Yes, she did./No, she didn't.
You had many books. [너는 많은 책을 가지고 있었다.]
→Did you have many books?–Yes, I did./No, I didn't.

Basic Test

※ ( ) 안에서 알맞은 말을 고르시오.

01 (Do, Does) your teacher like dogs?

02 (Do, Does) you play the violin?

03 (Do, Does) he drive a car?

04 (Do, Does) they study Chinese?

※ ( ) 안의 단어를 바르게 배열하시오.

05 (do, visit, you) your uncle every Saturday?

06 (Jennifer, did, see) her friend last night?

07 (they, live, do) in the beautiful house?

08 (the store, does, sell) pens?

※ 우리말과 뜻이 같도록 빈칸에 알맞은 말을 쓰시오.

09 당신은 차를 좋아합니까?
______ ______ like tea?

10 Jane은 매일 아침 운동을 하나요?
______ ______ exercise every morning?

11 Harry와 Sally는 과학을 공부합니까?
______ ______ study science?

※ 다음 의문문에 대한 대답을 완성하시오.

12 Does your brother watch TV every day?
–Yes, ______ ______ .

13 Does Sarah like rainy days?
–No,______ ______ .

Practice Test

01 다음 빈칸에 들어갈 알맞은 말을 고르시오.

A : Do you live in Seoul?
B : ________ I live in Incheon.

① Yes, I do. ② Yes, I don't.
③ Yes, I was. ④ No, I don't.
⑤ No, I'm not.

02 다음 빈칸에 들어갈 알맞은 말을 고르시오.

A : Do you need some help?
B : ________ Please pass me the salt.

① Yes, I am. ② No, you aren't.
③ Yes, I do. ④ No, I don't.
⑤ Yes, I don't.

03 다음 빈칸에 들어갈 알맞은 말을 고르시오.

A : ____________________
B : Yes, she does.

① Did she leave a message?
② Don't you make much money?
③ Didn't she study yesterday?
④ Does he have a Playstation at home?
⑤ Does she take the subway every morning?

04 다음 중 어법상 틀린 것을 고르시오.

① Do they know Mary?
② Did Mr. and Mrs. Smith fight again?
③ Didn't she take you to the park?
④ Did the parking lot full now?
⑤ Were you at the concert last evening?

05 다음 중 어법상 틀린 것을 고르시오.

① Do Tom and Mary love each other?
② Did you finish the work?
③ Didn't you stay at the hotel last night?
④ Do you usually walk to school or take the bus?
⑤ Does James show you his pictures last week?

06 다음 대화 중 어색한 것은?

① A : Do you play the violin every day?
B : No, I didn't.
② A : Did Tony win the game?
B : Yes, he did. It was exciting.
③ A : Did she come last night?
B : No, she didn't.
④ A : Does he have a digital camera?
B : Yes, he does. It's great.
⑤ A : Did you read The Da vinci Code?
B : Yes, I did. It's really fun.

Reading Comprehension

01 다음 글에서 밑줄 친 부분의 의미로 a. 의사가 의도한 것과 b. 남자가 이해한 것이 순서대로 바르게 짝지어진 것은?

> A man calls the hospital and gets a doctor on the line. He starts shouting on the phone, "Doctor, please, you have to help me. My wife is having a baby and I don't know what to do!"
> Doctor : Is this her first child?
> Man : No, you fool, this is her husband!

① a. 당신이 그녀의 첫 번째 남편입니까?
b. 이번이 첫 출산입니까?
② a. 그녀가 당신의 첫 번째 아이입니까?
b. 이번이 첫 전화입니까?
③ a. 이번이 첫 출산입니까?
b. 당신이 그녀의 첫 번째 남편입니까?
④ a. 이번이 첫 출산입니까?
b. 당신이 그녀의 첫 번째 아이입니까?
⑤ a. 그녀가 당신의 첫 번째 아내입니까?
b. 당신이 그녀의 첫 번째 남편입니까?

02 다음 글에서 노파가 왕자를 야수로 만든 이유로 알맞은 것은?

> Once upon a time, a handsome young prince lived in a big castle in a forest. One evening an old woman came to the castle. She asked the prince for a room to stay in during the night. When the prince said no, the old woman turned the prince into a beast. Then she gave him a rose and said, "This flower will live until your twenty first birthday. You and a girl must love each other before it dies. If not, you'll be a beast forever."
> 〈from Beauty and the Beast〉

① 온정을 베풀지 않아서
② 너무 사치스럽게 살아서
③ 왕의 명령을 듣지 않아서
④ 왕자의 외모를 시기해서
⑤ 진정한 사랑을 찾아주려고

명사와 관사

UNIT 1 명사의 단수와 복수

Prestudy

▣ 다음을 해석하시오.

1. She has two babies. They are good children.

____

2. She smiled with joy.

____

3. I want some milk.

____

▣ 다음을 영작하시오.

1. 나는 개 두 마리가 있다.

____

2. 그 토크쇼에 3명의 여성이 있었다.

____

3. 그는 아침에 커피 한 잔을 마신다.

____

Key Point

1. **명사** : 사람, 사물, 장소, 생각 등을 나타내는 말로 어떤 것의 '이름'을 말한다.

1) 셀 수 있는 명사 : 하나, 둘….로 셀 수 있는 명사를 의미한다. 단수인 경우 부정관사 a/an이 붙을 수 있고, 복수인 경우 -s를 붙이거나 불규칙하게 변하는 경우가 있다.

(1) 셀 수 있는 명사의 복수형 만드는 법

① 규칙 변화

대부분의 명사	명사+-s	maps, dogs, cats, bags, books
-s, -ss, -x, -ch, -sh, -o로 끝나는 명사	명사+-s	buses, kisses, boxes, watches, dishes, potatoes 〈예외〉pianos, photos
자음+y로 끝나는 명사	y를 i로 고치고+ -es	lady→ladies, baby→babies, city→cities
모음+y로 끝나는 명사	명사+ -s	boys, toys, monkeys
-f, -fe로 끝나는 명사	-f, -fe를 v로 고치고+ -es	leaf→leaves, knife→knives 〈예외〉roofs

② 불규칙 변화

ex) man→men woman→women foot→feet tooth→teeth goose→geese

단수와 복수가 같은 명사 : deer→deer(a deer, two deer) series→series sheep→sheep

2) 셀 수 없는 명사 : 하나, 둘,…로 셀 수 없는 명사를 의미한다. 단수형으로 쓰고 단수취급을 한다. 셀 수 없는 명사 앞에는 a/an을 쓰지 않는다.

① 추상명사 : joy, happiness, luck, love, beauty, news 등

② 고유명사 : Tom, busan, Korea 등(사람, 지역, 나라 이름 등) 고유명사의 첫 글자는 대문자로 쓴다.

③ 물질명사 : paper, water, milk, sugar, coffee, money 등

④ 셀 수 없는 명사의 수량 표현 : 단위나 용기 등을 나타내는 어구 사용

ex) a piece of paper/cheese/cake/advice, two glasses of water a cup of coffee, two bottles of beer

※ 항상 복수인 명사 : glasses(안경), scissors(가위), pants(바지)

복수처럼 보이는 단수명사 : news, physics

Basic Test

※ ( ) 안에서 알맞은 말을 고르시오.

01 There are two (box, boxes) in this room.

02 These (gentlemen, gentlemans) are from America.

03 Do the (ladies, ladys) go to church on Sundays?

04 I would like to have (two cup of coffee, two cups of coffee).

※ 다음 〈보기〉에서 알맞은 말을 골라 적절히 바꾸어 쓰시오. (단, 한 번씩만 쓰시오.)

보기	bottle, piece, glass, cup

05 Bring me a ______ of coffee.

06 He drank two ______ of water.

07 Here is a ______ of wine.

08 Jinhee bought a ________ of bread for lunch.

※ 다음 문장에서 어법상 틀린 부분을 찾아 바르게 고치시오.

09 Bring me two sheet of papers.

10 There are many potatos in the basket.

11 Some childs go to school by bicycle.

※ 다음 빈칸에 ( ) 안의 명사의 복수형을 쓰시오.

12 Ms. Baker has five ________: three ______ and two ______. (child, girl, boy)

13 Autumn is coming. The _____ are falling from the ______. (leaf, tree)

Practice Test

01 다음 중 복수형이 잘못 연결된 것은?

① potato–potatoes ② piano–pianos
③ sheep–sheeps ④ leaf–leaves
⑤ baby–babies

※ 다음 중 빈칸에 들어갈 수 없는 말을 고르시오.

02 There are many ______ in the water.

① fish ② child ③ swimmers
④ divers ⑤ boys

03 _____ has a beautiful flower in her hand.

① My wife ② My mother
③ A little girl ④ The tall women
⑤ Your sister

04 Yumin gave me a piece of ______ .

① milk ② cake ③ cheese
④ paper ⑤ advice

05 다음 중 어법상 틀린 것을 고르시오.

① I ordered two cups of coffee.
② She drank a glass of water.
③ Jack bought a pair of shoes.
④ This morning I drank a bottle of juices.
⑤ I want two slices of pizza.

06 다음 중 어법상 틀린 것을 고르시오.

① There are many pens in my bag.
② I always brush my teeth after meals.
③ There are many foxs on the farm.
④ There were many wolves in the zoo.
⑤ I saw many fallen leaves on the road.

07 다음 중 어법상 틀린 것을 고르시오.

① I burned my finger last Monday.
② Tomatoes are good for your health.
③ The actress has beauty and talent.
④ My school has eighteen classes.
⑤ People from America is quite tall.

08 다음 중 바르게 쓰인 문장은?

① He has two doges.
② Here is a pieces of cake.
③ There are two glasses of waters on the table.
④ Do you want a cup of coffees?
⑤ She smiled with joy.

Reading Comprehension

01 다음 글의 제목으로 가장 알맞은 것은?

At the age of 17, Korean pop singer BoA has it all—fame and money. What else could she want? BoA is highly popular in both Korea and Japan. In 2004 she received the MTV award. She is the only Asian pop artist to receive the same award twice in two years. BoA's first single came out in Japan in 2001. The next year, she climbed to the top of the J—pop charts with her fourth single, "Listen to My Heart." At the MTV awards ceremony, BoA said, " I want to succeed not only in Korea and Japan, but also in all of Asia."

① Asia's Most Famous Singer
② Recent MTV Award Winners
③ BoA's Life story
④ A Rising Korean Pop Star
⑤ Korea's Rich and Famous

02 다음 글에 나타난 필자의 심경 변화로 가장 알맞은 것은?

This morning I saw a man standing on the road. I shouted, but he didn't move. Suddenly a car came around the corner and hit him! I ran to him. The man was under the car! I pulled his arm, and it broke! I was holding it in my hand. Then I saw the reason—the man was a plastic model from the clothes shop.

① 즐거움→두려움
② 놀람→안심함
③ 따분함→흥미가 생김
④ 신남→화남
⑤ 부끄러움→슬픔

UNIT 2 관사

Prestudy

▣ 다음을 해석하시오.

1. It is a beautiful song.

2. I have a ball. The ball is in my room.

3. I go to school by bus.

▣ 다음을 영작하시오.

1. 그녀는 디자이너이다.

2. 나는 애완견을 한 마리 샀다. 그 개는 매우 귀여웠다.

3. 점심 식사 후에 여기서 만나자.

Key Point

1. **관사 :** 명사 앞에서 명사의 의미와 성격을 나타내는 말로 부정관사 a/an과 정관사 the가 있다.

1) 부정관사 a/an : 셀 수 있는 단수 명사 앞에 사용, 뒤에 나오는 단어의 발음이 자음으로 시작하면 a, 모음으로 시작하면 an을 사용

＊ 모음 [a], [e], [i], [o], [u]

ex) a student, an orange, an hour

① 막연한 하나를 나타낼 때 : He is a tourist.

② 하나(one)를 나타낼 때 : I have five English classes in a week.

③ ~마다, ~당(=per) : He goes jogging three times a week.

2) 정관사 the

① 앞에 나온 명사를 언급할 때 : We saw a movie. The movie was great.

② 서로 아는 것을 가리킬 때 : Open the door, please.

③ 명사가 수식어의 꾸밈을 받아 가리키는 대상이 분명할 때 : The key on the table is Bob's.

④ 관용적으로 쓰이는 경우 : play the piano(악기), the sea(자연환경), the sun(하나뿐인 존재), the Internet

3) 관사를 생략하는 경우 : before dinner(식사), play tennis(운동경기), by train(교통수단), go to school('공부하러 가다' 본래의 용도로 쓰일 때)

※ breakfast, lunch, dinner 등에는 관사를 쓰지 않는 것이 보통이나, 앞에 수식하는 말이 올 때는 부정관사 a/an을(를) 쓰기도 한다. Have a nice dinner.

Basic Test

※ (   ) 안에서 알맞은 말을 고르시오.

01 Do you have (a money, any money)?

02 He has (a, an) album.

03 Sunyoung likes to play (a, the) violin in her room.

04 The children always played (the baseball, baseball) together.

※ 다음 문장에서 어법상 틀린 부분을 찾아 바르게 고치시오.

05 He is a honest boy.

06 One day a boy was watching his sheeps near the woods.

07 Pass me a salt, please.

08 A sun is shining brightly in the sky.

※ 다음 빈칸에 알맞은 관사를 쓰시오. (불필요하면 ×표를 하시오.)

09 He plays ______ basketball every weekend.

10 How many seasons are there in ______ year?

11 Sujin keeps _____ cat and _____ dog. _____ cat is black and _____ dog is white.

※ 밑줄 친 부분에 유의하여 우리말로 옮기시오.

12 An apple a day keeps the doctor away.

____

13 Susan plays the guitar on the weekends.

____

Practice Test 

01 다음 빈칸에 들어갈 알맞은 말을 고르시오.

He has _____ cute smile.

① a ② an ③ the
④ those ⑤ 들어갈 말 없음

02 다음 빈칸에 들어갈 알맞은 말을 고르시오.

A : I play _____ violin in my free time.
B : Wow, that sounds interesting.

① a ② an ③ the
④ those ⑤ 들어갈 말 없음

03 다음 빈칸에 들어갈 알맞은 말을 고르시오.

A : Is Sangmin here?
B : No. He left here _____ ago.

① a hour ② an hour
③ two hour ④ an hours
⑤ the hour

04 다음 빈칸에 들어갈 말이 순서대로 바르게 짝지어진 것은?

I have _____ dog. _____ dog is very faithful.

① an–The ② a–The ③ a–A
④ an–An ⑤ the–The

05 다음 중 the가 필요 없는 문장은?

① Open _____ window, please.
② Mary traveled around _____ world.
③ _____ cat in the cage is mine.
④ In ____ afternoon, we always drink some tea.
⑤ Jin came home after _____ dinner.

06 다음 중 어법상 틀린 것은?

① She brought a cat. The cat was cute.
② She went to America by the plane.
③ Pete went to school in the morning.
④ Bring me the book on the shelf.
⑤ David plays basketball every weekend.

07 다음 우리말을 영어로 바르게 옮긴 것을 고르시오.

John은 한 달에 두 번씩 파티에 간다.

① John went to parties a month twice.
② John goes to parties twice the month.
③ John goes to parties is twice a month.
④ John goes to parties twice a month.
⑤ John went to parties twice the month.

Reading Comprehension

01 멜 깁슨에 관한 설명 중 다음 글의 내용과 일치하지 않은 것은?

> Mel Gibson is a famous Hollywood movie star. He is strong and handsome. People call him a tough guy. But he can be funny and shy, like a little boy. Mel Gibson has been in many movies such as Lethal Weapon, Braveheart and Ransom. He is now also a film director. He directed Braveheart, and it made him an Academy Award winner in 1995.

① 사람들은 그를 '터프 가이'라 부른다.
② 재미있고 수줍음이 많다.
③ Braveheart와 같은 영화에 출연했다.
④ Lethal Weapon으로 감독상을 수상했다.
⑤ 멜깁슨은 영화 감독이기도 하다.

02 다음 글의 주제로 알맞은 것은?

> Dogs and Cats are wonderful friends to people. They don't ask questions and complain like people. They just listen and seem to understand. They never get angry or fight with people. People with pets don't get so many colds and headaches. And when people play with cats or dogs, they forget about their troubles.

① 애완 동물의 장점
② 애완 동물 기르는 법
③ 애완 동물의 종류
④ 애완 동물의 습관
⑤ 애완 동물의 역사

의문사

› **UNIT 1.** who, what, which

› **UNIT 2.** when, where, why

› **UNIT 3.** how

UNIT 1 who, what, which

Prestudy

▣ 다음을 해석하시오.

1. Who is the man with sunglasses?

____

2. Which do you want?

____

3. What do you want?

____

▣ 다음을 영작하시오.

1. 저 멋진 남자는 누구니?

____

2. 네가 가장 좋아하는 TV 프로그램은 무엇이니?

____

3. 너는 액션 영화와 코미디 영화 중 어느 걸 더 좋아하니?

____

Key Point

1. **의문사 :** 누가, 언제, 어디서, 무엇을, 어떻게, 왜 등을 물을 때 사용하는 말
의문사가 있는 의문문에 대한 응답은 (Yes/No)로 대답(한다, 하지 않는다).

1) who : 누구
Who is she?−She is my classmates.
Who saw Jane?

(1) whoes : 누구의
Whose shoes are these?

(2) whome : 누구를
Who(m) do you like?

2) what : 무엇
What do you do after school?
What time do you go home?
* 의문형용사 what
What+명사 …? (what은 명사를 수식하는 의문형용사)
What time …?
What kind of+명사 …?

3) which : 어느 것
Which do you like better, pizza or chicken?
Which movie did you watch yesterday?

2. 의문사가 있는 의문문의 어순
1) 동사가 be동사일 때 : 의문사+be동사+주어 …?
2) 동사가 일반동사일 때 : 의문사+do(es)+주어+동사원형 …?
3) 의문사 자체가 주어일 때 : 의문사+동사 …?

Basic Test

※ 다음 빈칸에 알맞은 의문사를 쓰시오.

01 ______ is the largest city in the world?

02 ______ helped the poor child?–Minwoo did.

03 ______ do you like better, green tea or herbal tea?

04 ______ won the race?–James was the winner.

05 ______ is the matter with you?

06 다음 대화의 빈칸에 알맞은 의문사를 쓰시오.

A : ________ is her name?

B : Her name is Kim Sumi.

07 다음 대화의 빈칸에 알맞은 의문사를 쓰시오.

A : ________ music do you like, rock or jazz?

B : I like jazz.

08 다음 대화의 빈칸에 알맞은 의문사를 쓰시오.

A : ________ is that in your hand?

B : It's an eraser.

09 다음 대화의 빈칸에 알맞은 의문사를 쓰시오.

A : ______ is she over there?

B : She is my English teacher, Ms. Park.

※ 주어진 대답에 알맞은 질문을 〈보기〉에서 고르시오.

보기

① Who's this in the picture?
② What time is it?
③ What does your father do?
④ Who is your father?
⑤ Who is the man?
⑥ What do you want?

10 A : ______________________

B : It's my mother.

11 A : ______________________

B : My father is a businessman.

12 A : ______________________

B : I want some juice.

13 A : ______________________

B : It's seven.

Practice Test

01 다음 빈칸에 들어갈 알맞은 말을 고르시오.

A : _____ is that boy with sunglasses?
B : That's Mike.

① Which ② What ③ Whom
④ Who ⑤ Why

02 다음 빈칸에 들어갈 알맞은 말을 고르시오.

A : _____ does Mrs. Taylor do?
B : She's a cook.

① Which ② What ③ Whom
④ Who ⑤ Why

03 다음 빈칸에 들어갈 알맞은 말을 고르시오.

A : _____ do you like better, coffee or tea?
B : I like tea better.

① Which ② What ③ Whom
④ Who ⑤ Why

04 다음 우리말을 영어로 바르게 옮긴 것은?

파티에서 누가 Dana와 춤을 췄니?

① Who dances with Dana at the party?
② Which did dance with Dana at the party?
③ Who danced with Dana at the party?
④ Dana danced with who at the party?
⑤ What danced with Dana at the party?

05 다음과 같은 대답이 나올 수 있는 질문은?

A : ____________________
B : I watched a new movie.

① What do you do?
② Who did saw you?
③ Which movie would you like to see?
④ What did you do after school?
⑤ What kind of movies do you like?

06 밑줄 친 질문에 대한 답이 될 수 <u>없는</u> 것은?

I saw you at the park yesterday.
<u>Who was with you?</u>

① I went there with Jisu.
② Oh, I was with Jisu.
③ I was alone.
④ No, I didn't see you.
⑤ Jisu was with me.

07 다음 중 어법상 <u>틀린</u> 것은?

① Who fought with him?
② Whose building is that?
③ Whom was that girl?
④ Who stood outside the door?
⑤ Whom do you like best among them?

Reading Comprehension

01 다음 글에 나타난 필자의 심경 변화로 알맞은 것은?

Last year my aunt and uncle invited me to visit them in America. On the day of my trip, I went to the airport. But I arrived there a few minutes late and missed the plane. I was standing in the airport thinking "What an idiot I am!" when I heard the news : The plane had crashed! I couldn't believe it. I was very lucky to be alive!

① 들뜸→실망함
② 지루함→놀람
③ 불안함→슬픔
④ 놀람→화가 남
⑤ 안타까움→안도함

02 다음 중 빈 칸에 알맞은 것은?

Some trees change their colors in the fall. Their green leaves turn red or yellow. Then the leaves drop. Other trees stay green all year. They don't lose their leaves in the fall. The old leaves stay on until new ones grow. These are evergreen trees. The word ever means "always" and the trees always ____________ .

① turn yellow
② stay green
③ have old leaves
④ drop their leaves
⑤ change their colors

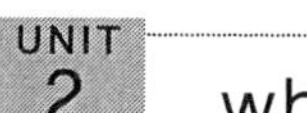

when, where, why

Prestudy

▣ 다음을 해석하시오.

1. When do you have dinner?

____

2. Where is my bag?

____

3. Why are you so happy?

____

▣ 다음을 영작하시오.

1. 그의 생일은 언제니?

____

2. 근처에 우체국이 어디에 있어요?

____

3. 너는 왜 너의 여자 친구를 좋아하니?

____

Key Point

1. when : 언제

A : When does the concert start?

B : It starts at 9 p.m.

A : When is your birthday?

B : My birthday is August 12th.

* 시간을 묻는 when은 what time으로 바꿔 쓸 수 있다.

When did you buy a digital camera?=What time did you buy a digital camera?

2. where : 어디서, 어디에

A : Where did you get this coupon?

B : I bought it in Internet.

A : Where is book store?

B : Go straight and turn right at the corner.

3. why : 왜

A : Why were you late for school?

B : Because I got up late.

* Why에 대한 대답은 주로 Because으로 시작한다.

1) 〈Why don't you … ?〉 ~하는 것이 어때? 상대방에게 권유할 때 쓰는 표현

Why don't you take a break? You look tired.

2) 〈Why don't we … ?〉 우리 ~하지 않을래? 함께하자는 권유의 표현

Why don't we go swimming?

Basic Test

01 ( ) 안에서 알맞은 말을 고르시오.

(Where, When) are you, Mom?−I'm in the kitchen.

02 ( ) 안에서 알맞은 말을 고르시오.

(Why, When) do you swim in the river?−In August.

03 ( ) 안에서 알맞은 말을 고르시오.

(Where, When) do you get up?−I get up at six.

04 ( ) 안에서 알맞은 말을 고르시오.

(When, Why) are you late today?−Because of a car accident.

05 ( ) 안에서 알맞은 말을 고르시오.

(Where, Who) is your father?−He is in the yard.

※ 다음 질문에 대한 대답으로 알맞은 것을 〈보기〉에서 고르시오.

보기
① Because I was sick yesterday.
② I studied English in England.
③ I'm from Egypt.
④ In 2002.

06 Alice : Where are you from?
Jack : ______________________

07 Kate : Why were you absent from school?
Sam : ______________________

08 Paul : When did you travel to Europe?
Youngsoo : ______________________

09 다음 빈칸에 공통으로 들어갈 알맞은 말을 쓰시오.

________ do you live? ________ is Miss Chris from?

10 다음 빈칸에 공통으로 들어갈 알맞은 말을 쓰시오.

_______ did you meet her first? _______ is your birthday?

11 우리말과 뜻이 같도록 주어진 단어를 배열하시오.

왜 그가 파티에 오지 않았지? (he, didn't, to, the party, come, why)

______________________________?

Practice Test 

01 다음 빈칸에 들어갈 알맞은 말을 고르시오.

A : _____ were you born?
B : In 1993.

① What ② Where
③ When ④ Why
⑤ How

02 다음 빈칸에 들어갈 알맞은 말을 고르시오.

A : _____ didn't you do your homework yet?
B : Because I was so busy all week.

① What ② Where
③ When ④ Why
⑤ How

03 다음 빈칸에 들어갈 알맞은 말을 고르시오.

A : Where did Tom go last year?
B : ____________________

① At 10 o'clock.
② His last name is Jones.
③ He's Clara's English teacher.
④ He went to Australia.
⑤ He's too young.

04 다음 우리말을 영어로 바르게 옮긴 것은?

다음 영어 시험이 언제인가요?

① When was the next English exam?
② What time does the next English exam?
③ When is the next English exam?
④ Where is the next English exam?
⑤ Why does the next English exam?

※ 다음 질문에 대한 답이 될 수 없는 것을 고르시오.

05 When does the concert start?

① At noon.
② It starts at 7.
③ It starts in an hour.
④ It already began.
⑤ It's a fantastic concert.

06 Why don't you take a little break?

① Sounds good.
② I'm too busy now.
③ I don't break anything.
④ That's a great idea.
⑤ I'm OK.

Reading Comprehension

01 다음 글의 밑줄 친 부분에 대한 이유로 알맞은 것은?

A deaf couple checked into a motel. In the middle of the night, the man went out to get some aspirin. He found the bottle of aspirin in the car. Then he realized something : he couldn't remember his room number! He thought and thought and finally he got an idea. He opened the car door again and started blowing the horn. <u>Soon the people turned their lights on in every room-except for one window</u>.

① 아내가 외출했기 때문에
② 아내가 잠이 들었기 때문에
③ 방의 전등이 고장났기 때문에
④ 아내의 귀가 들리지 않기 때문에
⑤ 모텔 주인이 전등을 끄라고 했기 때문에

02 다음 글의 목적으로 알맞은 것은?

We have been playing at Pigeon Park since we were very little. It is a much better place to play than anywhere else. In the Summer we swim in the pool. In the Fall we play football. In the Winter we make snowmen and go sledding. In the Spring we fly kites. Please don't let the the Nord Company put up a building there. Please save Pigeon Park.

Sincerely
John and Tommy

① 피전 공원 내 행사 초대
② 피전 공원의 놀이 시설 광고
③ 피전 공원 내 건물 설립 방지 촉구
④ 피전 공원의 환경 오염 실태 보고
⑤ 피전 공원 내 놀이 기능 강화 촉구

UNIT 3 how

Prestudy

▣ 다음을 해석하시오.

1. How do you make bread?

____

2. How tall is he?

____

3. How long is the movie?

____

▣ 다음을 영작하시오.

1. 캐나다 날씨는 어떠니?

____

2. 너는 그녀의 이름을 어떻게 알았니?

____

3. 너는 펜이 몇 개 필요해?

____

Key Point

1. how : 어떤, 어떻게

A : How do you go to work?

B : by bus.

A : How are you today?

B : I'm great.

2. how+형용사/부사

① How old…? 나이를 물을 때 사용. 몇 살이니?

How old is he?

② How long …? 얼마나 오랫동안, 얼마나 긴

How long does it take to get to the station?

③ How often… ? 얼마나 자주

How often do you go to hair salon?

④ How many+셀 수 있는 명사 (복수형) …?

How much+셀 수 없는 명사 …?

How many notebooks do you need?

How much gas do you want?

⑤ How high~ 얼마나 높은가?

How high is Mt. Everest?

⑥ How far~ 얼마나 먼가?

How far is it from Busan to Seoul?

⑦ How fast~ 얼마나 빠른가?

How fast is the train?

⑧ How deep~ 얼마나 깊은가?

How deep is the river?

⑨ How wide~ 얼마나 넓은가?

How wide is the room?

* How[What] about+명사/v-ing …? ~은 어때? ~하는 것은 어때? (권유)

Basic Test

※ 다음 두 문장의 빈칸에 공통으로 들어갈 알맞은 말을 쓰시오.

01 ______ do you go to school?
______ much money do you have in your pocket?

02 ______ do you do after school?
______ time is it?

03 ______ much is this CD player?
______ many hamburgers do you want?

※ 다음 밑줄 친 부분을 어법에 맞게 고치시오.

04 How <u>many</u> sugar do you need?

05 <u>What</u> do you go to school?–I go to school by subway.

06 <u>Who</u> are you today?–Good, thank you.

07 How <u>much</u> books are there on the desk?

※ 주어진 대답에 알맞은 질문을 〈보기〉에서 고르시오.

보기
① Do you read many books?
② How much is the skirt?
③ How old is your brother?
④ How do you go to work?
⑤ How many books do you read in a year?
⑥ Do you go to school by bus?

08 A : ______________________
B : He is seven years old.

09 A : ______________________
B : It is $20.

10 A : ______________________
B : Ten books a year.

11 A : ______________________
B : By bus.

Practice Test

01 다음 빈칸에 들어갈 알맞은 말을 고르시오.

A : _____ is your vacation?
B : It's two months.

① How long ② How old
③ How often ④ How many
⑤ How

02 다음 빈칸에 들어갈 알맞은 말을 고르시오.

A : _____ does he come here?
B : Twice a week.

① How long ② How old
③ How often ④ How many
⑤ How

03 다음 빈칸에 들어갈 알맞은 말을 고르시오.

A : _____ is the shirt?
B : It's 30,000 won.

① What ② How much
③ When ④ How many
⑤ Where

04 다음 우리말을 영어로 바르게 옮긴 것은?

중국엔 얼마나 많은 사람들이 있죠?

① How are people there in China?
② How much are there in China?
③ How old are there in China?
④ How many people are there in China?
⑤ How much people are there in China?

05 다음과 같은 대답이 나올 수 있는 질문은?

A : ______________
B : It was great.

① How old are you?
② How often do you travel?
③ How was your trip to Guam?
④ How many people are there?
⑤ What about going on a trip?

06 다음 말과 의미가 다른 것은?

How about going to Everland tomorrow?

① Let's go to Everland tomorrow.
② What about going to Everland tomorrow?
③ Why don't we go to Everland tomorrow?
④ Do you like to go to Everland tomorrow?
⑤ Shall we go to Everland tomorrow?

Reading Comprehension

01 다음 글에서 밑줄 친 to go의 의미로 a. 여자가 의도한 바와 b. 필자가 이해한 바가 바르게 짝지어진 것은?

There are many differences in the meaning of English words. But I didn't know that when I first came to America. One day I ordered my food at McDonald's. A girl at the counter asked me "to go?" I was angry and said, "I came here to eat. Why should I go?" She explained for a few minutes. Then I understood "to go" means to take the food away and not to eat there.

① a. 자리에 가 계시겠어요?
   b. 기다리시오.
② a. 주문하고 가시겠어요?
   b. 떠나시오.
③ a. 가지고 가실건가요?
   b. 가시오.
④ a. 배달을 원하세요?
   b. 가시오.
⑤ a. 여기서 드실 겁니까?
   b. 기다리시오.

02 다음 글이 주는 교훈으로 알맞은 것은?

Jim Abbott was born without a right hand. Like many boys, Jim loved to play baseball. His dream was to become a major league pitcher. But how could he play with one hand? How would he be able to throw and catch?

Jim's dad told him that he should believe in himself. He showed him a special way to catch and throw with his left hand. Jim practiced day and night. At last his dream came true. In 1989, Jim began playing in the major leagues!

① 정직한 사람만이 성공할 수 있다.
② 노력하면 어떤 일이든 해낼 수 있다.
③ 자신에 대한 반성이 성공의 열쇠가 된다.
④ 꿈을 이루어 냈다고 자만해서는 안 된다.
⑤ 최고가 된다고 반드시 행복 한 것은 아니다.

동사의 시제

› **UNIT 1.** 현재시제와 과거시제

› **UNIT 2.** 진행시제

UNIT 1 현재시제와 과거시제

Prestudy

▣ 다음을 해석하시오.

1. She eats breakfast at 7 every day.

____

2. Plants need water.

____

3. I was in the hospital a month ago.

____

▣ 다음을 영작하시오.

1. 그는 패스트푸드를 먹지 않는다.

____

2. 나는 어제 그와 함께 즐거운 시간을 보냈다.

____

Key Point

1. **현재시제**

1) 현재의 성질/상태, 일상적인 습관/동작, 불변의 진리/사실 등을 나타낼 때

He feels happy. (현재의 감정/생각)

I go jogging in the park every morning. (현재의 일상 습관)

Two and five is seven. (불변의 진리)

2. **과거시제**

1) 이미 끝난 과거의 동작/상태 등을 나타낼 때

Kate went to the museum last Saturday. (과거동작)

I was on TV yesterday. (과거의 상태)

2) 과거시제는 흔히 과거를 나타내는 시간의 부사(구)와 함께 쓰인다.

yesterday, last night[week, month, year, monday, ago 등

3) 역사적 사실은 과거시제로 쓴다.

The Korean War broke out in 1950.

3. **현재완료** : have/has+P.P(과거분사)

1) 과거의 사실이 현재까지 영향을 미치는 경우

(1) 경험 : ~한 적이 있다.

I have visited Jejudo twice.

(2) 완료 : 현재까지 동작이 완료된 것

She has already read the newspaper.

(3) 계속 : 과거부터 지금까지 계속 ~해 왔다.

I have been here for two hours.

(4) 결과 : 과거의 사건이 현재까지 영향을 미치는 경우

She has lost the book.

① 단순과거 : 과거에 샀다는 사실만 서술

I bought a computer.

현재완료 : 현재도 컴퓨터를 가지고 있다는 의미

I have bought a computer.

② 단순과거 : 밥을 먹은 사실만 의미

I had breakfast.

현재완료 : 현재도 배가 고프지 않다는 의미

I have had breakfast.

Basic Test

※ ( ) 안에서 알맞은 말을 고르시오.

01 David (goes, went) to London last week.

02 He (begins, began) swimming lessons one month ago.

03 Last year I (traveled, travel) to China.

04 Did you go to the party last night?–No, I (stayed, stay) at home.

※ 우리말과 뜻이 같도록 ( ) 안의 말을 알맞게 바꾸어 쓰시오.

05 그들은 지난 여름 바다에서 수영을 했다. (swim)

They _____ in the sea last summer.

06 그녀는 작년에 미국에서 한국으로 돌아왔다. (come)

She _______ back to Korea from America last year.

07 나는 매일 아침 식사 후 신문을 읽는다. (read)

I _____ the newspaper after breakfast every day.

08 아버지께서 작년에 MP3 플레이어를 사주셨다. (buy)

My father _____ an MP3 player last year.

※ 다음 밑줄 친 부분을 바르게 고치시오.

09 Gandhi <u>dies</u> in 1948.

10 The sun <u>rose</u> in the east every day.

11 King Sejong <u>invents</u> hangeul.

12 Last year I <u>learn</u> Spanish in Spain.

13 In 1936 Son Gijeong <u>wins</u> a gold medal in a marathon.

Practice Test

※ 다음 빈칸에 들어갈 알맞은 말을 고르시오.

01 Jennie ____ to the office every Sunday morning.

① come ② comes
③ came ④ comed
⑤ is coming

02 She read Harry Potter ______.

① now ② tomorrow
③ next Monday ④ last night
⑤ next weekend

03 Jake ___ a train to London last year.

① take ② took
③ taking ④ is taking
⑤ takes

04 다음 밑줄 친 부분 중 어법상 틀린 것을 고르시오.

① Mong didn't have a girlfriend last year.
② The Korean War broke out in 1950.
③ They bought an LCD TV yesterday.
④ Morris has six cats at home now.
⑤ Heejin travels to Europe 3 years ago.

05 다음 밑줄 친 부분 중 어법상 틀린 것을 고르시오.

① Peter feels happy now.
② He goes home by subway every day.
③ They invented a new board game in 2004.
④ The sun rises in the east.
⑤ We watched movies on Saturdays.

06 다음 우리말을 영어로 바르게 옮긴 것은?

Brian은 가을에 자주 등산을 간다.

① Brian often go mountain climbing in the fall.
② Brian often went climbing in the fall.
③ Brian often goes mountain climbing in the fall.
④ Brian often climbed in the fall.
⑤ Brian often go mountain climb in the fall.

07 다음 밑줄 친 단어의 알맞은 형태는?

Sam : Where did you stay 2 years ago?
Sally : I stay in Canada.

① stayed ② stay
③ stays ④ am staying
⑤ staied

Reading Comprehension

01 다음 글의 내용과 일치하지 <u>않는</u> 것은?

To travel in space is dangerous. So chimpanzees were sent into space before men! In the 1950s, scientists sent chimps into space because they are much like humans. The first chimps to ride in a rocket were Pat and Mike. It was in 1953. But Ham was the first chimpanzee to go into space. That was in 1961.

① 침팬지들이 사람보다 먼저 우주여행을 했다.
② 침팬지가 우주 여행을 하게 된 것은 인간과 닮았기 때문이다.
③ 최초로 로켓에 탄 침팬지는 팻과 마이크였다.
④ 팻과 마이크가 최초로 우주 여행을 한 침팬지이다.
⑤ 침팬지의 최초 우주 여행은 1961년에 이루어졌다.

02 다음 글의 분위기로 알맞은 것은?

Three old men are at the doctor's office to take a memory test. The doctor says to the first old man, "What is three times three?" "274" he answers. The doctor says to the second man. "It's your turn. What is three times three?" "Tuesday," the second man replies. The doctor says to the third man, "OK, What's three times three?" "Nine," says the third man. "That's great," says the doctor. "How did you get that?" "Simple," says the third man. "I took 274 away from Tuesday."

① 비장하다.
② 으스스하다.
③ 긴박하다.
④ 권태롭다.
⑤ 유머러스하다.

UNIT 2 진행시제

Prestudy

▣ 다음을 해석하시오.

1. I'm washing my dog.

___

2. They weren't playing football.

___

3. Was he running then.

___

▣ 다음을 영작하시오.

1. 내 남자친구가 무대 위에서 춤추고 있다.

___

2. Jim은 그의 블로그를 만들고 있었다.

___

3. 너 내 말 듣고 있니?

___

Key Point

1. **진행시제** : 특정 시점에서 진행 중인 일을 묘사할 때

1) 현재 진행시제 : be동사(am/are/is)+v–ing(~하고 있다.)
She is baking cookies.

2) 과거 진행시제 : be동사(was/were)+v–ing(~하고 있었다.)
They were riding a roller coaster.

2. 진행시제의 부정문과 의문문

1) 부정문 : be동사+not+v–ing( )
I'm not listening to music.

2) 의문문 : Be동사+주어+v–ing( )
Were they watching TV at that time?

3. 진행형(v–ing) 만드는 법

대부분의 동사	동사원형+–ing	play→playing
–e로 끝나는 동사	e를 빼고+–ing	skate→skating
–ie로 끝나는 동사	ie를 y로 고치고+–ing	lie→lying
〈단모음+단자음〉으로 끝나고 그 단모음에 강세가 오는 동사	자음을 한 번 더 쓰고+–ing	begin→beginning
1음절 동사 중 〈단모음+단자음〉으로 끝나는 동사		stop→stopping

Basic Test

01 ( ) 안에서 알맞은 말을 고르시오.

I (am, was) taking a bath last night.

02 ( ) 안에서 알맞은 말을 고르시오.

I (take, was taking) a walk yesterday.

03 ( ) 안에서 알맞은 말을 고르시오.

Kim (is cooking, cook) in the kitchen.

04 ( ) 안에서 알맞은 말을 고르시오.

The baby (was, is) sleeping soundly yesterday.

05 다음 밑줄 친 부분을 바르게 고치시오.

Last night Jieun were cooking in her kitchen.

06 다음 밑줄 친 부분을 바르게 고치시오.

Minsu and his sister was washing the dishes yesterday.

07 다음 밑줄 친 부분을 바르게 고치시오.

Last night they are reading comic books.

08 다음 밑줄 친 부분을 바르게 고치시오.

BoA is stayed in Japan.

※ 주어진 대답에 알맞은 질문을 〈보기〉에서 고르시오.

보기
① What was he doing yesterday?
② Where are you going?
③ What is he doing?
④ Where were you going yesterday?

09 A : ______________________
B : I'm going home.

10 A : ______________________
B : He's doing his homework.

11 A : ______________________
B : We were going to the market.

※ 밑줄 친 부분에 유의하여 우리말로 옮기시오.

12 They were singing a song in the classroom.

____

13 She has read *The Lord of the Rings* three times.

____

New Words The Lord of the Rings 반지의 제왕

Practice Test

01 다음 중 동사와 〈v-ing〉형이 잘못 연결된 것은?

① read−reading ② play−playing
③ cut−cutting ④ lie−liing
⑤ shake−shaking

※ 다음 빈칸에 들어갈 알맞은 말을 고르시오.

02 He was ____ a bicycle near the seaside.

① ride ② rides ③ rode
④ rided ⑤ riding

03 Was she ____ the house?

① clean ② cleaning
③ cleaned ④ cleans
⑤ will clean

04 Yunho ____ on the Internet last night.

① are chatting ② aren't chatting
③ was not chatting ④ not is chatting
⑤ is not chatting

05 다음 우리말을 영어로 바르게 옮긴 것을 고르시오.

그는 어젯밤에 컴퓨터를 수리하고 있었다.

① He fixed the computer last night.
② He is not fixing the computer last night.
③ He was fixing the computer last night.
④ He was fixed the computer last night.
⑤ He's fixing the computer last night.

06 다음 우리말을 영어로 바르게 옮긴 것을 고르시오.

Tom은 음악 파일을 다운로드 받고 있다.

① Tom is downloading some music files.
② Tom downloads some music files.
③ Tom downloaded some music files.
④ Tom was downloading some music files.
⑤ Tom is download some music files.

07 다음 밑줄 친 부분 중 어법상 틀린 것을 고르시오.

① Terry was crying last night.
② Taeji was taking her sister to school that morning.
③ Subin's takes care of her pets.
④ He's singing a famous French song on stage.
⑤ Melinda was walking downstairs slowly then.

Reading Comprehension

01 다음 글의 내용상 빈 칸에 가장 알맞은 것은?

There were some very ______ laws in the past.

- In the eighteenth century in England, people had to pay a "window tax" for each window in their house. So poor people often built their houses without windows.
- In the nineteenth century, women teachers in the U.S.A. couldn't get married or even date. If they got married, they had to give up their job.

① good
② fair
③ strange
④ effective
⑤ useful

※ 다음 글을 읽고 물음에 답하시오.

Some fish lay their eggs in the water and leave them. They do not wait until the egg hatch. ______ tilapias don't leave their eggs. The father keeps all the eggs safe in its mouth. It keeps them there for twenty days. Then the eggs hatch, and the baby fish swim away. <u>So the poor father can't eat for twenty days!</u>

02 윗 글의 밑줄친 So the poor father can't eat for twenty days!의 원인으로 알맞은 것은?

① 겨울잠을 자기 때문에
② 알을 입에 물고 있기 때문에
③ 먹이를 구하기가 어렵기 때문에
④ 알을 지키느라고 힘이 빠졌기 때문에
⑤ 다른 물고기들의 공격을 받았기 때문에

03 윗 글의 빈 칸에 알맞은 것은?

① But
② Therefore
③ Although
④ And
⑤ Because

조동사

UNIT 1 will, be going to

Prestudy

▣ 다음을 해석하시오.

1. We'll read the magazine together.

____

2. I won't borrow money from you.

____

3. She's going to forgive him.

____

▣ 다음을 영작하시오.

1. 나는 거기에 너와 함께 가지 않을 것이다.

____

2. 회의는 오후 3시에 시작될 것이다.

____

3. 우리는 네가 그리울 거야.(be going to)

____

Key Point

1. **조동사 :** 조동사는 동사를 도와 의미를 완성해 주는 것이다.
동사에 미래, 가능, 추측, 허락, 부탁 등의 의미를 더해 주는 말

1) Will+동사원형 : ~할 것이다(미래), ~하겠다(주어의 의지)

It will rain tomorrow. (미래)

I won't(will not) change my mind. (의지)

−Will you join the sports club? (의문문의 어순 : 조동사+주어+동사원형~?)

−No, I won't.

(1) 조동사 will의 과거형 : would

① 과거 시점에서 미래를 나타낼 때 : He said that he would meet her.

② 정중한 부탁을 나타낼 때 : Would you close the door?

③ would like to+동사원형 ~하고 싶다 : I would like to go on a diet.

(2) 주어+will 의 줄임말 : 주어+'ll

I'll, You'll, She'll, He'll, It'll, We'll, They'll

(3) will+not=won't

dy read the newspaper.

2) be going to+동사원형 : ~할 것이다(가까운 미래), ~할 예정이다(계획)

I'm going to travel to Europe this summer.

He's not going to stay up all night. (be going to의 부정 : be not going to)

−Are you going to call her? (의문문의 어순 : Be동사+주어+going to+동사원형)

−Yes, I am.

Basic Test

※ ( ) 안에서 알맞은 말을 고르시오.

01 I (will, am going) play baseball after school today.

02 Jack (is going, will) come to Seoul in October.

03 She (wills, is going to) go to Harvard University next year.

04 He will (buy, buys) a sports car.

05 We (will not go, will go not) there tomorrow.

※ 다음 대화의 빈칸에 알맞은 말을 쓰시오.

06 A : I feel cold. Will you close the door for me?
B : Yes, I ______________.

07 A : Will you read this book?
B : No, I __________. I will read The Lord of the Rings.

※ 우리말과 뜻이 같도록 ( ) 안의 단어를 배열하시오.

08 나는 윤도현 콘서트에 가지 않을 거야. (to, go, not, will)
I _____ the Yoon Dohyeon concert.

09 이번 주말에 뭐 할 거야? (weekend, going, to, you, are, do, this)
What ______________ ?

10 내일은 날씨가 좋을 거야. (tomorrow, be, will, fine)
It ______________.

※ 밑줄 친 부분에 유의하여 우리말로 옮기시오.

11 He <u>will</u> become a good artist.
____

12 We <u>will</u> go to the movies on Saturday.
____

13 How long are you <u>going to</u> stay here?
____

Practice Test

※ 다음 빈칸에 들어갈 알맞은 말을 고르시오.

01 Tony _____ to join us tomorrow morning.

① will be ② be going ③ going
④ are going ⑤ is going

02 Larry _____ after the baby next week.

① looks ② will look
③ look ④ was looking
⑤ will looks

03 Harry _________ back to Hogwarts this winter.

① will come ② come
③ will comes ④ came
⑤ am going to

04 다음 빈칸에 들어갈 수 <u>없는</u> 것은?

I will wait for you at the main gate _____.

① tomorrow morning
② tonight
③ this evening
④ last night
⑤ the day after tomorrow

05 다음 우리말을 영어로 바르게 옮긴 것은?

내일 코엑스몰에 갈 거니?

① Will you going to COEX Mall tomorrow?
② Were you going to COEX Mall tomorrow?
③ Will you go to COEX Mall tomorrow?
④ Do you going to COEX Mall tomorrow?
⑤ Did you go to COEX Mall tomorrow?

06 다음 중 어법상 <u>틀린</u> 것을 고르시오.

① Will you write a memo for me?
② When are you going to leave?
③ I'm going to talk to you last night.
④ Would you open the door?
⑤ Is Sarah going to visit us next Tuesday?

07 다음 중 어법상 <u>틀린</u> 것을 고르시오.

① He's not going to play this game.
② Toby not will tell you the story.
③ She won't play the flute tomorrow.
④ Aren't you going to join the meeting?
⑤ They're not going to play video games.

Reading Comprehension

01 다음은 무엇에 대한 글인가?

I'll never forget the first time I went to a football match. The match was between our local team and a team from a nearby city. When we walked through the gate of the stadium, I was surprised to see so many people. When the game started, it was really exciting. People shouted the whole time. And when our team scored, they got excited and cheered loudly. Unfortunately, we lost the match, but I had a great time and I became a real football fan.

① 처음으로 축구장에서 시합한 일
② 축구 경기를 처음으로 직접 본 일
③ 처음으로 축구 선수의 사인을 받은 일
④ 처음으로 우리 지역 축구팀이 승리한 일
⑤ 축구 응원 단장이 되어 처음으로 관중 앞에 선 일

※ 다음 글을 읽고 물음에 답하시오.

Keiko is the whale which was in the movie Free Willy. He has been in a large fish tank for years. Keiko's owners want to set him free. But they are not sure. Can he talk with other whales? Will he be able to hunt? Right now Keiko does not seem ready. (그는 한 번도 바다에 혼자 나가 본적이 없다.) He is spending 15 hours a day with humans. He still eats frozen fish. He needs humans too much. He should not go into the sea alone for some time.

02 윗 글에 Keiko에 대한 설명으로 틀린 것은?

① 영화 '프리 윌리'에 출연한 고래이다.
② 수년간 큰 수족관에서 지냈다.
③ 주인은 그를 풀어주고 싶어 한다.
④ 주로 살아 있는 물고기를 잡아 먹는다.
⑤ 사람들의 도움을 많이 필요로 한다.

03 윗 글에 밑줄 친 부분을 영어로 옮길 때 빈 칸에 알맞은 것은?

He ______________ in the ocean alone.

① have never be
② have never been
③ has never be
④ has never been
⑤ has never was

UNIT 2 can, may

Prestudy

▣ 다음을 해석하시오.

1. Mina can run fast.

____

2. You can see my album.

____

3. You may use the phone.

____

4. He may be a middle school student.

____

▣ 다음을 영작하시오.

1. 나는 피자를 만들 수 있다.

____

2. 제가 도와드릴까요?

____

3. 오늘 저녁에 비가 오리지도 모른다.

____

Key Point

1. Can

1) 능력/가능 ~할 수 있다.(= be able to)

He can answer the question. =He is able to answer the question.

Can you swim?－Yes, I can./No, I can't.

2) 허락 ~해도 좋다.(= may)

Can I use your cell phone －Yes, you can.

You can't go there alone.

3) 조동사 can의 과거형 : could

① 과거시점에서 능력을 나타낼 때

I thought Korean team could win the game.

② 공손한 부탁을 나타낼 때

Could you pass me the salt?

4) 확신의 can

must(~임에 틀림없다)의 부정은 cannot(~일 리가 없다)

He cannot be a singer.

2. may

1) 허락 ~해도 좋다.(can보다 정중한 표현)

You may eat dessert after dinner.

May I use this coupon?－Yes, you may./No, you may not.

2) 추측 ~일지도 모른다.

He may arrive soon.

The rumor may not be true.

Basic Test

※ ( ) 안에서 알맞은 말을 고르시오.

01 Jang Nara is able (to sing, sing) very well.

02 May I use your MP3 player?—Yes, you (may, do).

03 I am very busy. I (can't go, can go not) to the movies tonight.

04 (Can, May) you swim?—Yes, I can.

※ 우리말과 뜻이 같도록 ( ) 안의 단어를 배열하시오.

05 그는 다리를 다쳤다. 그래서 그는 잘 걸을 수가 없다. (walk, very, cannot, well)

He hurt his leg. So he ______________.

06 숙제를 다 마쳤니? 그러면 나가서 농구해도 된다. (go, out, basketball, play, and, may)

Did you finish your homework? Then you ______________________________.

07 그는 체스를 둘 수 있다. (chess, play, is, to, able)

He ______________________________.

※ 다음 밑줄 친 부분을 바르게 고치시오.

08 Betty <u>able</u> to speak Chinese.

09 Sally can <u>cooks</u> very well.

10 May I <u>helping</u> you?

※ 밑줄 친 부분에 유의하여 우리말로 옮기시오.

11 Mark speaks Korean a little. He <u>may</u> be learning Korean.

____

12 Sarah Jang <u>can</u> play the violin very well.

____

13 <u>Could</u> you help me, please?

____

Practice Test

01 다음 빈칸에 들어갈 알맞은 말을 고르시오.

A : I'm hungry. _____ I have some bread?
B : Yes, you may.

① Do ② Am ③ May
④ Will ⑤ Is going to

02 다음 빈칸에 들어갈 알맞은 말을 고르시오.

A : May I use your phone?
B : ____________________

① Sure. Here it is. ② Yes, you may not.
③ Yes, I may. ④ No, you can.
⑤ I'm not sure.

03 다음 빈칸에 들어갈 알맞은 말을 고르시오.

A : Can you play the violin?
B : _____, but I will someday.

① Yes, I can ② Yes, I can't
③ No, I can ④ No, I can't
⑤ I'm sorry

04 다음 중 may의 쓰임이 나머지 넷과 <u>다른</u> 것은?

① It <u>may</u> be Mrs. Robinson.
② <u>May</u> I sit here?
③ She <u>may</u> be very rich.
④ They <u>may</u> be in the building.
⑤ That <u>may</u> be your mistake.

05 다음 중 can의 쓰임이 나머지 넷과 <u>다른</u> 것은?

① She <u>can</u> play baseball.
② <u>Can</u> you speak Chinese?
③ Danny <u>can't</u> dance at all.
④ You <u>can't</u> swim in this pond.
⑤ I <u>can't</u> keep a diary in English.

06 다음 중 어법상 <u>틀린</u> 것을 고르시오.

① You cannot go out this weekend.
② Can you fix this computer?
③ May I use your pen?
④ He may is rich.
⑤ Can you come to the party?

07 다음 중 어법상 <u>틀린</u> 것을 고르시오.

① The rumor may be true.
② Can you inline skate well?
③ He may join our club.
④ Myungsu may be with Sumi at the park.
⑤ Ben may have not the MP3 player.

Reading Comprehension

01 다음 글에 밑줄 친 This building에 대한 설명 중 바르지 못한 것은?

> <u>This building</u> is the original main building of the Lord and Lady Department Store Company. It was designed by the famous architect Archibald Morgan. It was built by the Vanderpool Construction Company. Construction was begun in 1845 and was finished in 1847. Since its opening day, the Lord and Lady Department Store has been thought as one of the finest examples of 19th century American building. Nowadays the store is visited the presidents and famous people of many countries.

① 백화점 건물이다.
② 설립자는 Archibald Morgan 이다.
③ 2년 만에 세워졌다.
④ 19세기 미국 건축물이다.
⑤ 유명 인사들이 방문하는 곳이다.

※ 다음 글을 읽고 물음에 답하시오.

> At the airport, a woman was checking in a long line of travelers. Suddenly an angry passenger pushed his way to the front. He said, " I'm a FIRST CLASS passenger and I HAVE TO be on this flight." The woman answered, "I'm sorry, sir. You have to wait your turn." The passenger then said very loudly, "Listen, do you know who I am?" Without an answer, the woman smiled and picked up a microphone. "<u>May I have your attention please? We have a passenger here. He does not know who he is. If anyone can help him, please come to the gate.</u>" With a red face, the man finally walked to the back of the line.

02 윗 글에서 여자가 밑줄 친 부분과 같이 말한 이유는?

① 남자 승객을 돕기 위해
② 남자 승객의 가족을 찾기 위해
③ 남자 승객이 자기가 누구인지 몰라서
④ 남자 승객이 일등석에 탄다는 것을 알리기 위해
⑤ 남자 승객의 잘못된 태도를 우회적으로 지적하려고

UNIT 3 must, have to, should

Prestudy

▣ 다음을 해석하시오.

1. You must see the doctor.

____

2. We don't have to rush.

____

3. You should search websites.

____

▣ 다음을 영작하시오.

1. 너는 이 일을 해결해야 한다.

____

2. 그는 건강을 위해 살을 빼야 한다. (have to)

____

3. 너는 잔디를 밟으면 안 된다. (should)

____

Key Point

1. must

1) 의무(~해야 한다.=have to)

You must finish this report.

You must not tell a lie.

2) 강한 추측(~임에 틀림없다.)

She must be a doctor.

He cannot be here now; he's in New York.

* 반대말 : cannot (~일 리가 없다)

3) must의 부정

① 금지 : 해서는 안된다.

You must not go there now.

② 추측 : ~일 리가 없다.

It cannot be true.

2. have to : 의무(~해야 한다. =must)

Tom has to go home before dark.

Do we have to leave now? −No, we don't have to.

* have to의 부정 : don't have to (~할 필요가 없다)

3. should : 의무, 충고, 제안(~해야 한다.)

You should eat more. You're skinny.

You shouldn't break your promise.

* must보다 강제성이 약한 의무 또는 권고를 나타낸다.

You should rest. (=You had better rest.)

※ 〈had+better+동사원형〉 : ~하는 것이 낫다.

(이것은 'had'가 과거 형태이나 뜻은 현재시제이다.)

A : I have a cold.

B : You had better take some medicine.

※ 의무의 강제성 : must〉have to〉should

Basic Test

※ ( ) 안에서 알맞은 말을 고르시오.

01 Do you want to go to university? Then you (must, may) pass the exam.

02 A : Smith is absent from school.
B : Oh, he (must, should) be sick.

03 A : Do I have to solve the question?
B : No, you (don't have to, don't must).

04 A : Can you help me?
B : Sorry, I cannot help you. I (must, can) do my homework first.

※ 다음 밑줄 친 부분을 바르게 고치시오.

05 Must he cleans the room?

06 He musts help his father.

07 We should watch not too much TV.

08 A : I have a cold.
B : You should to see the doctor.

※ 우리말과 뜻이 같도록 ( ) 안의 단어를 배열하시오.

09 이 셔츠는 너무 작아. 나는 그것을 교환해야 해. (it, exchange, must)
This shirt is too small.
I ______________________.

10 그는 너무 많이 먹으면 안 된다. (should, eat, too, not, much)
He ______________________.

※ 밑줄 친 부분에 유의하여 우리말로 옮기시오.

11 It's raining hard. You should stay at home.
____

12 Bill Gates owns Microsoft. He must be rich.
____

13 You don't have to come on Sunday.
____

New Words own 소유하다

Practice Test

01 다음 우리말을 영어로 옮길 때, 빈칸에 알맞은 말은?

잔디밭 위를 걸으면 안 된다.
→ You ________ walk on the grass.

① must ② not should
③ must not ④ should
⑤ don't have to

02 다음 빈칸에 들어갈 알맞은 말을 고르시오.

A : I'm feeling sick.
B : Hey, you ________ to take a rest.

① can ② will
③ must ④ have
⑤ should

03 다음 빈칸에 들어갈 알맞은 말을 고르시오.

A : Do I have to turn off the computer now?
B : No, ________. I'll use it in a minute.

① you have to ② you won't
③ you doesn't ④ you can't
⑤ you don't have to

04 다음 밑줄 친 부분 중 어법상 틀린 것을 고르시오.

① You'd better be at home by 9.
② David must stay not in the room.
③ He has to see her parents.
④ She doesn't have to clean the house.
⑤ You should bring your dictionary to every class.

05 다음 밑줄 친 부분 중 어법상 틀린 것을 고르시오.

① She must be a rich girl.
② Joan should run to school right now.
③ She have to work.
④ Tommy must take the book to the office.
⑤ Jenny has to make tonight's dinner.

06 다음 중 must의 쓰임이 나머지 넷과 다른 것은?

① She must be very pretty.
② They must call the police now.
③ You must come back home by tomorrow.
④ You must read these novels by next week.
⑤ You must not smoke in this building.

Reading Comprehension

01 앤이 브라운 선생님의 조언을 듣고 친구에게 했을 말로 알맞은 것은?

> Dear Ms. Brown,
>
> I've got a problem with one of my friends. I really like her, but she always copies me. If I buy new clothes, the next day she will buy the same ones. It's the same with music, videos, even my friends! It really bothers me. What should I do?
>
> Ann
>
> Dear Ann,
>
> Tell her that you don't like it when she copies you. She may not be very sure of herself. So she may want to be like you. You should talk to her about your feelings.
>
> Ms. Brown

① "I feel sad that we're not friends any more."

② "It's time you let me borrow some of your clothes."

③ "My friend doesn't like you because you copy her."

④ "Do you know that you always copy me? Sorry, but I don't like it."

⑤ "Why do you use my things? Please don't wear my clothes."

02 고대 이집트의 고양이에 관한 설명 중, 다음 글의 내용과 일치하지 않는 것은?

> In ancient Egypt cats were loved and worshiped because they protected the grain storehouses from mice. They were even protected by Egyptian law. To harm a cat was a serious crime, and you could be killed for killing a cat. When a cat died, all the family members became sad. They shaved off their eyebrows as a sign of their sadness. The dead cat was then honored and buried in a serious ceremony.

① 쥐로부터 곡식을 지켰다.

② 법의 보호를 받았다.

③ 죽인 사람은 사형을 당할 수도 있었다.

④ 죽으면 눈썹이 잘렸다.

⑤ 죽은 후에도 소중하게 다뤄졌다.

문장의 형태

› **UNIT 1.** 주어+동사+(수식어)

› **UNIT 2.** 감각동사+형용사

› **UNIT 3.** 동사+목적어

› **UNIT 4.** 수여동사+간접목적어+직접목적어

› **UNIT 5.** 동사+목적어+목적보어

■ 문장의 요소와 품사

1. 영어 단어는 8가지의 품사로 나뉘고 그 품사에 따라 문장에서의 역할을 한다.

They work hard in the field every day. [그들은 매일 들판에서 열심히 일한다.]

* 주어인 대명사 They, 동사 work, 부사인 hard, 전치사구인 in the field, 시간 부사인 every day로 이루어진 문장이다.)

2. 품사와 문장에서의 쓰임

쓰임	역할	example
명사	주어, 목적어, 보어	The dog ate the meat. [그 개는 고기를 먹었다.]
대명사	주어, 목적어, 보어	He loves her very much. [그는 그녀를 매우 많이 사랑한다.]
동사	동사	They play basketball in the gym. [그들은 체육관에서 농구를 한다.]
형용사	보어, 명사수식	She is a beautiful lady. [그녀는 아름다운 숙녀이다.]
부사	동사, 형용사, 부사, 문장 수식	She can speak Korean very well. [그녀는 한국어를 매우 잘 말한다.]
접속사	연결	He is a teacher and I am a doctor. [그는 선생님이고 나는 의사이다.]
전치사	전치사구를 이루어 형용사나 부사의 역할	She gets up early in the morning. [그녀는 아침에 일찍 일어난다.]
감탄사	감탄을 나타낸다.	Oh, it is very cute. [오! 그것은 매우 귀엽다.]

3. 문장의 요소

요소	역할	example
주요소	주어	She calls me Jen. [그녀는 나를 Jen이라고 부른다.] (주어 She, 동사 calls, 목적어 me, 보어 Jen)
	목적어	
	보어	
	동사	
부속요소	형용사적 수식어	She is a smart student. [그녀는 영리한 학생이다.] (smart는 명사인 student를 수식한다.)
	부사적 수식어	Tom runs fast. [Tom은 빨리 달린다.] (부사인 fast는 동사 runs를 수식한다.)

UNIT 1 주어+동사+(수식어)

Prestudy

▣ 다음을 해석하시오.

1. I go there every year.

____

2. There is a book in the drawer.

____

3. She gets up early in the morning.

____

▣ 다음을 영작하시오.

1. 탁자 위에 꽃병이 하나 있다.

____

2. 여기 버스가 온다.

____

3. 그는 8시에 학교에 간다.

____

Key Point

1. **1형식 문장 :** 1형식 문장은 [주어+동사+(수식어)]로 이루어진 문장을 말하고 1형식 문장을 만드는 동사를 완전자동사라고 한다. 완전자동사는 목적어와 보어 모두 사용하지 않는 동사이다.

   The boys ran fast in the park. [그 소년들은 공원에서 빨리 달렸다.]

   * 주어인 The boys, 동사인 ran, 그리고 부사인 fast와 장소 부사구인 in the park로 이루어진 문장이다. 문장의 주요소로는 주어와 동사가 있는 것이다.)

2. **완전자동사의 도치**

쓰임	형태	example
There	There+동사+주어	Some books are on the table. →There are some books on the table. [탁자 위에 몇 권의 책이 있다.]
Here	Here+동사+주어	The bus comes. →Here comes the bus. [버스가 여기 온다.]
장소, 방향 부사구	장소, 방향 부사구+동사+주어	The sun rises in the east. →In the east rises the sun. [해는 동쪽에서 뜬다.]

3. **There로 도치될 때의 유의점 :** There로 주어와 동사를 도치하는 경우 주어가 새로운 것이므로 강조를 하기위한 것이다. 그러므로 There로 시작하는 문장에서 주어에는 the를 붙일 수가 없다.

   There is the book on the desk. (×)

   →There is a book on the desk. (○)

4. **Here로 도치될 때의 유의점 :** Here로 주어와 동사를 도치하는 경우 주어가 인칭대명사일 경우는 안 된다. 그러므로 주어가 인칭대명사일 때는 Here+주어+동사의 형태로 사용한다.

   Here are you.(×)

   →Here you are.(○)

Basic Test

01 다음 문장 중 어법상 어색한 것은?

① There is a vase on the table.

② There is a book on the desk.

③ There lived a pretty girl in the house.

④ There used to be an old oak tree.

⑤ There was the boy in the room.

02 다음 문장 중 어법상 어색한 것은?

① Here we are.

② In the east of Asia Korea lies.

③ To the north marched the army.

④ On the table was the book.

⑤ There lived a wise king in the country.

03 다음 문장의 빈칸에 들어갈 알맞은 것은?

> 교실 안에는 많은 소년들이 있었다.
> → _____ _____ a lot of boys in the classroom.

① There are　　② Here are

③ There were　　④ Here were

⑤ There is

04 다음 문장을 영어로 바르게 옮긴 것은?

> 달은 서쪽으로 진다.

① The moon set in the west.

② In the west the moon set.

③ The moon sets west.

④ West the moon sets.

⑤ In the west sets the moon.

05 다음 문장 중 문장의 형식이 다른 것은?

① Some flowers are in the vase.

② There were five girls in the class.

③ The old lady lived in the country with her husband.

④ Tom and Jane are good friends.

⑤ They went to church last Sunday.

06 다음 문장과 의미가 같은 것은?

> The machine is on the table.

① The table the machine is on.

② Table the machine is on the.

③ On the table the machine is.

④ On the table was the machine.

⑤ On the table is the machine.

07 다음 문장 중 어법상 어색한 것은?

① There are boys in the classroom.

② He lived in the city with her.

③ Birds sang merrily in the tree.

④ To the south Tom went.

⑤ Here come Tony and Jane.

Practice Test

01 다음 문장 중 어법상 <u>어색한</u> 것은?

① There were a few vases on the table.
② There were some books on the desk.
③ There lived a wise king in the country.
④ There used to be a pond in the village.
⑤ There is the book in the drawer.

02 다음 문장 중 어법상 <u>어색한</u> 것은?

① Here comes the train.
② In the east rises the sun.
③ To the north went Tom and Jane.
④ There is a pencil in the pencil case.
⑤ There are a student over there.

03 다음 문장의 빈칸에 들어갈 알맞은 것은?

> 체육관에 많은 학생들이 있다.
> → _____ _____ a lot of students in the gym.

① There are ② Here are
③ There were ④ Here were
⑤ There is

04 다음 문장을 영어로 바르게 옮긴 것은?

> 탁자위에 그 장난감이 있었다.

① There is the toy on the table.
② There was the toy on the table.
③ On the table was the toy.
④ On the table the toy was.
⑤ On the table is the toy.

05 다음 문장 중 문장의 형식이 <u>다른</u> 것은?

① A lot of books are in the library.
② There are a few boys in the class.
③ There lived an old farmer in the country.
④ He came back home early yesterday.
⑤ They were smart enough to solve the problem.

06 다음 문장과 의미가 같은 것은?

> A pond used to be in the village.

① Here is a pond in the village.
② In the village a pond used to be.
③ There is a pond in the village.
④ There used to be a pond in the village.
⑤ They had a pond in the village.

Reading Comprehension

01 다음 글에서 예견한 미래의 모습과 다른 것은?

In the future, we probably won't buy food and cook it. We'll simply choose instant meals. And food won't be bought at the supermarket. We'll select meals from a menu on our computer screens. The food will be delivered. And we will just heat it and eat it. This food won't be liked by everybody. But it'll give us more free time.

① 식품 구입을 위한 슈퍼마켓 이용이 감소한다.
② 컴퓨터로 음식을 주문한다.
③ 배달된 음식을 데우기만 하면 된다.
④ 음식 요리 시간을 많이 줄일 수 있다.
⑤ 느긋하게 식사를 즐기는 문화가 유행한다.

※ 다음 글을 읽고 물음에 답하시오.

Do you know how a moth is different from a butterfly? A butterfly has knobs on its antenna. A butterfly is active during the daytime. A butterfly has brighter colors than a moth. A butterfly folds its wings upward when it rests. A butterfly has a smooth body, and a moth may have a hairy one. Although almost everyone likes the butterfly, many people don't like the moth.

02 이 글의 주된 내용을 가장 잘 나타낸 문장으로 가장 적절한 것은?

① Many people don't like the moth.
② A moth is different from a butterfly.
③ A butterfly has knobs on its antenna.
④ A butterfly has brighter colors than a moth.
⑤ A butterfly has a smooth body, and a moth has a hairy one.

03 이글의 내용과 일치하지 않는 것은?

① 나비는 더듬이 끝에 둥근 마디가 있다.
② 나비는 낮 시간에 활동을 한다.
③ 나비는 나방보다 밝은 색을 띤다.
④ 나비는 쉴 때 날개를 위로 접는다.
⑤ 나비와 나방의 몸통은 털이 많다.

New Words
moth 나방
knob 둥근 마디

UNIT 2 감각동사+형용사

Prestudy

▣ 다음을 해석하시오.

1. She felt sick.

____

2. The fish doesn't smell good.

____

3. He is an engineer.

____

▣ 다음을 영작하시오.

1. 그녀는 어려 보인다.

____

2. 이 신선한 과일은 맛이 좋다.

____

3. 그는 우리를 만나러 달려서 왔다.

____

Key Point

1. **2형식 문장** : 2형식 문장은 [주어+동사+보어]로 이루어진 문장을 말한다. 보어는 명사나 형용사에 해당하는 것들이 사용된다. 이때 사용되는 동사를 불완전자동사라고 한다.
   Jane is a middle school student. [Jane은 중학생이다.]
   * 주어인 Jane, 동사인 is, 보어인 a middle school student로 이루어진 문장이다.

2. 불완전 자동사의 분류

분류	종류	example
be 류	be, stay, keep 등	You should stay healthy for the next match. [너는 다음 시합을 위해 건강한 상태를 유지해야 한다.]
become 류	become, get, grow, turn, go, come 등	The players grew tired. [선수들은 피곤해 졌다.]
표정동사	seem, appear 등	The box seems heavy to carry. [그 상자는 나르기에 무거워 보인다.]
감각동사	look, sound, smell, taste, feel	That sounds strange. [그것은 이상하게 들린다.]

3. 감각동사의 쓰임

종류	형태	example
look sound smell feel taste	주어+동사+형용사	He looks happy. [그는 행복해 보인다.] (우리말로는 부사처럼 해석이 되지만 영어로는 형용사를 사용한다는 것에 주의해야 한다.)

4. 감각동사+like

종류	형태	example
look sound smell feel taste	주어+동사+like+명사	He looks like a teacher. [그는 선생님처럼 보인다.] (like는 전치사로 '~처럼'이라는 뜻으로 사용된다. 그러므로 명사가 전치사의 목적어로 온다.)

Basic Test

※ 우리말과 뜻이 같도록 빈칸에 알맞은 말을 쓰시오.

01 그 해산물은 맛이 신선하지 않았다.

The seafood didn't taste ______________.

02 너의 영어 선생님은 키가 커 보이신다.

Your English teacher looks ____________.

03 그의 손은 차가웠다.

His hands felt ________________________.

04 그녀의 목소리는 부드럽게 들린다.

Her voice sounds ______________________.

※ 다음 밑줄 친 부분을 바르게 고치시오.

05 Her baby looks <u>love</u>.

06 My piano didn't sound <u>well</u>.

07 The game looked <u>excitingly</u>.

※ 우리말과 뜻이 같도록 빈칸에 알맞은 말을 쓰시오.

08 이 종이는 촉감이 거칠다.

This paper ____________________ rough.

09 그 농구팀은 아주 강해 보였다.

The basketball team __________ very strong.

10 이 피자는 맛있다.

This pizza __________ delicious.

※ 밑줄 친 부분에 유의하여 우리말로 옮기시오.

11 Jisu <u>looks sick</u>.

____

12 The soup <u>tastes salty</u>.

____

13 The carpet in your living room <u>feels</u> soft.

____

Practice Test

01 다음 문장 중 어법상 어색한 것은?

① The leaves turned red and yellow.

② Jenny became famous in her country.

③ Eric felt boring during the holiday.

④ He should keep it fresh.

⑤ He looked happy after he won the game.

02 다음 문장 중 어법상 어색한 것은?

① He dies a begger.

② The dog fell a prey to the tiger.

③ She seems to be happy.

④ He appeared tired in the evening.

⑤ The theory proved falsely.

03 다음 문장의 빈칸에 들어갈 알맞은 것은?

> 그 뉴스는 나에게 이상하게 들린다.
> → The news sounds ______ to me.

① strange ② strangely

③ to strange ④ strangeness

⑤ more strangely

04 다음 문장 중 어법상 어색한 것은?

① His hair turned gray when he was in forties.

② Her face turned pale when she heard the news.

③ He looked like an expert.

④ The food tasted like pizza.

⑤ She looked like lovely.

05 다음 문장의 빈칸에 들어갈 수 없는 것은?

> She looked like ____________ .

① a teacher ② kind

③ a nurse ④ a doctor

⑤ an engineer

06 다음 문장의 빈칸에 들어갈 알맞은 것은?

> He looked ____________ .

① happy ② happily

③ a professor ④ a stranger

⑤ sorrowfully

07 다음 문장 중 어법상 어색한 것은?

① The bread tastes good.

② Jenny looks good today.

③ She looks like a lawyer.

④ The news sounds strangely to me.

⑤ The leather feels soft.

Reading Comprehension

01 다음 글의 내용과 일치하지 <u>않는</u> 것은?

Dear Sue,

I'm in the hospital! I've broken my foot! But don't worry. I'll be all right. I've been here since last Sunday. I had an accident when I was playing football. I was trying to kick the ball, but I kicked the goal post! The nurses and doctors have been really nice. I may go home tomorrow. I hope so–it's a bit boring here.

All the best Mark

PS : I've just heard! I am going home the day after tomorrow.

① 마크는 지난 일요일에 병원에 입원했다.
② 마크는 축구를 하다가 발이 부러졌다.
③ 마크는 상대팀 선수의 발에 걸려 넘어졌다.
④ 간호사와 의사들이 친절했다.
⑤ 마크는 병원 생활이 지겹다.

※ 다음 글을 읽고 물음에 답하시오.

Manuel had stomach cancer. After chemotherapy, Manuel's hair began to fall out. Soon he had no hair. He was sad. He felt strange without hair. He didn't want people to see him. One day his friends visited Manuel in the hospital. He was surprised when he saw them. They had no hair! He began to laugh. They laughed, too. When Manuel came home from the hospital , other friends with no hair came to visit him. They said, "We want to look like you." Now he is not sad any more.

02 Manuel의 심정의 변화로 가장 적절한 것은?

① bad→afraid ② sad→happy
③ surprised→sad ④ happy→surprised
⑤ excited→lonely

03 이 글이 시사 하는 바로 가장 적절한 것은?

① 서로 같은 부류끼리 모인다.
② 백짓장도 맞들면 낫다.
③ 어려울 때 친구가 진정한 친구이다.
④ 사귀는 친구를 보면 그 사람을 안다.
⑤ 천리 길도 한 걸음부터.

New Words chemotherapy 화학 요법

UNIT 3 동사+목적어

Prestudy

▣ 다음을 해석하시오.

1. I dreamed a strange dream.

____

2. I think that he is honest.

____

3. They played basketball after school.

____

▣ 다음을 영작하시오.

1. 나는 너를 다시 보기를 바란다.

____

2. 나는 그녀가 예쁘다고 생각한다.

____

3. 그는 맛있는 피자를 먹었다.

____

Key Point

1. **3형식 문장** : 3형식 문장은 [주어+동사+목적어]의 형태로 이루어진 문장을 말한다. 목적어는 동작의 대상이 되는 것을 말하며 명사에 해당하는 것, 즉 명사, 대명사, 부정사, 동명사, 재귀대명사, 명사절이 사용된다.
   She made some cookies yesterday. [그녀는 어제 약간의 쿠키를 만들었다.]
   * (주어인 She, 동사인 made, 목적어인 some cookies, 시간부사인 yesterday로 이루어진 문장이다.)

2. 3형식 문장의 분류

분류	의미	example
부정사를 목적어로 가지는 동사	want, hope, wish, expect, decide, refuse 등	He expected to pass the test. [그는 시험에 통과할 것이라고 기대했다.]
동명사를 목적어로 가지는 동사	enjoy, finish, mind, avoid, give up 등	She finished doing the dishes. [그녀는 설거지를 끝냈다.]
명사절을 목적어로 가지는 동사	think, believe, say 등	They think that she is very smart. [그들은 그녀가 매우 영리하다고 생각한다.]

3. 자동사로 착각하기 쉬운 3형식 동사

표현	의미	example
marry	결혼하다	She married a diplomat. [그녀는 외교관과 결혼했다.]
enter	들어가다	He entered the room. [그는 방에 들어갔다.]
discuss	토론하다	They discussed the matter. [그들은 그 문제를 토론했다.]
resemble	닮다	She resembles her mother. [그녀는 엄마를 닮았다.]
reach	도착하다	He reached the station. [그는 역에 도착했다.]

이 동사들은 뒤에 전치사를 사용하지 않도록 해야 한다.

She married with a diplomat.(×)

Basic Test

01 다음 문장 중 어법상 잘못된 것은?

① He likes all kinds of sports.
② They wanted to go there with her.
③ She enjoyed to meet them in the gallery.
④ The thief entered the room through the window.
⑤ I believe that everything is possible.

02 다음 문장 중 어법상 잘못된 것은?

① Mary married Tom last year.
② She entered the classroom.
③ They discussed about the problem with the teacher.
④ He resembles his father very closely.
⑤ They will reach Seoul tomorrow morning.

03 다음 문장의 빈칸에 들어갈 알맞은 것은?

그녀는 성공하기를 바란다.
→She hopes ___________ .

① succeed ② to succeed
③ to success ④ succeeding
⑤ to succeeding

04 다음 문장의 빈칸에 들어갈 알맞은 것은?

창문을 열어도 될까요?
→Would you mind _____ the window?

① open ② to open
③ opening ④ to opening
⑤ opened

05 다음 문장 중 어법상 잘못된 것은?

① He wants to be a famous actor.
② Mike expected to meet her at the station.
③ They say that she is very beautiful.
④ John didn't give up trying to solve the problem.
⑤ She finished to do her homework.

06 다음 문장의 빈칸에 들어갈 알맞은 것은?

나는 그녀가 매우 현명하다고 믿었다.
→I believed ___________ she was very wise.

① what ② that
③ if ④ whether
⑤ so

Practice Test

01 다음 문장 중 어법상 <u>잘못된</u> 것은?

① Tom liked all kinds of sports.

② He wants to go fishing with his friends.

③ They enjoyed eating delicious food in the restaurant.

④ He entered into the house through the door.

⑤ They believe that he can do the work.

02 다음 문장 중 어법상 <u>잘못된</u> 것은?

① He will marry Mary next year.

② The students entered the classroom.

③ Tom discussed the trouble with his mother.

④ Mike resembles after his mother.

⑤ He reached Seoul yesterday.

03 다음 문장의 빈칸에 들어갈 알맞은 것은?

> Tom은 그 일을 하기로 결심했다.
> →Tom decided ________ the work.

① do ② doing
③ to do ④ did
⑤ done

04 다음 문장의 빈칸에 들어갈 알맞은 것은?

> 그는 그녀를 만나는 것을 피했다.
> →He avoided ________ her.

① meet ② to meet
③ meeting ④ to meeting
⑤ met

05 다음 문장 중 어법상 <u>잘못된</u> 것은?

① Tony wanted to be a singer.

② Brian expects to get the gold medal.

③ They think that he is honest and kind.

④ He didn't mind giving me the book.

⑤ She enjoyed to singing English songs.

06 다음 문장의 빈칸에 들어갈 알맞은 것은?

> 그들은 그가 결백하다고 말했다.
> →They said ________ he was innocent.

① what ② that
③ if ④ whether
⑤ to

Reading Comprehension

01 Susan B. Anthony가 사람들에게 한 말의 요지로 알맞은 것은?

Susan B. Anthony was born in the U.S. in 1820. At that time, only men could vote. She thought, "Why can't women vote? We must change the law." She talked about this to people all around the country. In 1872, she voted, but her vote wasn't counted. And she was put in prison. When she died in 1906, women could vote only in four states. Today every woman in America can vote.

① "선거구를 재조정하라."
② "여성에게도 투표권을 달라."
③ "죄수들의 인권을 보장하라."
④ "투표를 통해 의원을 뽑아라."
⑤ "남성 위주의 사고 방식을 고쳐라."

※ 다음 글을 읽고 물음에 답하시오.

Puppies need a lot of ___________ when they are first born. They can't even see. Puppies don't open their eyes until they are 10 to 14 days old. They drink their mother's milk until they are 4 to 5 weeks old. Then they are ready for puppy food. Puppies like to stay with their mothers and brothers and sisters for a while. They should not be taken home until they are about 8 weeks old.

02 Puppies에 관한 설명중 글의 내용과 일치하는 것은?

① 처음 태어나자마자 눈을 뜬다.
② 생후 14일 동안은 어미 젖을 먹는다.
③ 생후 4~5주가 지나면 먹이를 먹는다.
④ 8개월 동안은 어미와 함께 있어야 한다.
⑤ 8주 동안 어미와 떨어지는 적응기를 갖는다.

03 빈 칸에 알맞은 말은?

① care　② milk
③ food　④ family
⑤ light

UNIT 4 수여동사+간접목적어+직접목적어

Prestudy

▣ 다음을 해석하시오.

1. He brought the students new books.

____

2. She made me a sandwich.

____

3. Mother cooked them a nice dinner.

____

▣ 다음을 영작하시오.

1. James는 우리에게 영어를 가르친다.

____

2. 나는 내 사진을 그녀에게 보여 주었다.

____

3. 나에게 사과주스를 좀 가져다 줄 수 있겠니?

____

Key Point

1. 4형식 문장 : 4형식 문장은 [주어+동사+간접목적어+직접목적어]로 이루어진 문장이다. 4형식을 만드는 동사를 수여동사라고 하고 간접목적어와 직접목적어는 서로 다른 것이 된다.
   She gave me a book yesterday. [그녀는 어제 나에게 책 한 권을 주었다.]
   * 주어인 She, 동사인 gave, 간접목적어인 me, 직접목적어인 a book으로 이루어졌다.

2. 4형식 동사의 분류

분류	사용하는 전치사	example
give, lend, pay, send, tell, show, teach, bring, sell, write 등	to	He gave her a blue box. →He gave a blue box to her. [그는 그녀에게 파란색 상자를 주었다.]
make, buy, cook, get, find 등	for	I bought my father a tie. →I bought a tie for my father. [나는 아빠에게 넥타이를 사 주었다.]
ask	of	He asked me a difficult question. →He asked a difficult question of me. [그는 나에게 어려운 질문을 했다.]

3. **직접목적어가 대명사일 때** : 3형식 문장의 형태로 사용해야 한다. 영어에서 새로운 정보와 이미 알고 있는 정보가 있을 때 새로운 정보가 뒤에 오는 경향이 있다. 그러므로 대명사로 사용된 직접목적어는 이미 알고 있는 정보가 되어서 앞에 오는 것이다.
   He gave me it.(×)
   →He gave it to me.(○)
   [그는 그것을 나에게 주었다.]

Basic Test

※ 우리말과 뜻이 같도록 빈칸에 알맞은 말을 쓰시오.

01 어제 아빠가 나에게 1,000원을 주셨다.

My father ________ me a thousand won yesterday.

02 그의 할머니가 우리들에게 피자를 사주셨다.

His grandmother ____________ us pizza.

03 Nancy가 Tom에게 선물을 줄 것이다.

Nancy will ________ Tom a present.

※ 두 문장의 뜻이 같도록 빈칸을 채우시오.

04 Mr. Baker teaches math to us.

→Mr. Baker teaches ______________.

05 I wrote an email to my friend.

→I wrote ______________________.

06 She made some cookies for her children.

→She made ____________________.

07 ( ) 안의 단어를 알맞은 곳에 넣으시오.

My ① brother ② lent ③ a ④ CD ⑤. (me)

08 ( ) 안의 단어를 알맞은 곳에 넣으시오.

The ① students ② asked ③ a ④ question ⑤. (their teacher)

09 ( ) 안의 단어를 알맞은 곳에 넣으시오.

My father ① bought ② a ③ digital ④ camera ⑤. (me)

※ 우리말과 뜻이 같도록 주어진 단어를 배열하시오.

10 삼촌이 나에게 DVD 플레이어를 주셨다. (a DVD player, gave, me)

My uncle ______________________.

11 그녀의 남편은 그녀에게 점심을 만들어 주었다. (made, for, lunch, her)

Her husband ____________________.

12 그는 그의 아들에게 스케이트 한 켤레를 사 줄 것이다. (his son, a pair of skates, buy)

He will ________________________.

13 버스 정류장으로 가는 길 좀 알려 주시겠습니까? (show, the bus stop, the way, to, me)

Will you _______________________?

Practice Test

※ 다음 빈칸에 들어갈 알맞은 말을 고르시오.

01 He bought a mirror __________ me.

① at ② to ③ for
④ of ⑤ in

02 She sent a fan letter __________ Lee Jungi.

① at ② to ③ for
④ of ⑤ in

03 My friend, Danny, asked a question__________ me.

① of ② to ③ for
④ at ⑤ in

04 다음 중 어법상 틀린 것을 고르시오.

① My father bought me a nice car.
② Jeremy gave a white shirt to me.
③ Can you bring me a red pen?
④ Hyunyoung told me to a truth.
⑤ My English teacher asked me many ques tions.

05 다음 중 어법상 틀린 것을 고르시오.

① I can't lend this MP3 player to you.
② She taught me English last year.
③ Please show me something new.
④ Jinhee showed her cat for me.
⑤ Naomi got me a new cell phone.

06 다음 우리말을 영어로 바르게 옮긴 것은?

> 그녀의 어머니는 그녀에게 노란 스웨터를 만들어 주셨다.

① Her mother made for her a yellow sweater.
② Her mother made a yellow sweater her.
③ Her mother made a yellow sweater to her.
④ Her mother made a yellow sweater of me.
⑤ Her mother made her a yellow sweater.

07 다음 우리말과 뜻이 같도록 할 때, 빈칸에 들어갈 말이 순서대로 짝지어진 것을 고르시오.

> She'll _____ Dotoris _____ her boyfriend.
> (그녀는 남자친구에게 도토리를 사줄 것이다.

① buys−to ② bought−to
③ bought−for ④ buy−for
⑤ buy−of

Reading Comprehension

01 다음 글의 내용과 일치하지 않은 것은?

> In 1985, ten short films were shown in Paris for the first time. Sixteen years later, the film industry was born in Hollywood. The first films were silent. The words were written at the bottom of the screen. In 1927, Warner Brothers in Hollywood made the first talking film, The Jazz Singer. It was a great success.

① 10편의 단편 영화가 처음 상연된 곳은 파리이다.
② 영화 산업이 시작된 곳은 할리우드이다.
③ 최초의 영화들은 무성영화였다.
④ 최초의 무성영화를 만든 영화사는 워너 브라더스 사이다.
⑤ 영화 The Jazz Singer는 대성공을 거두었다.

※ 다음 글을 읽고 물음에 답하시오.

> A man went to the hospital. The doctor said, "Well, Mr. Jones, you're going to have some shots, and then you'll feel much better." Then a nurse came to Mr. Jones. She said, "I'm going to give you your first shot now. Where do you want it?" The man was surprised and said, "Do you really want me to choose now?" "Yes, Mr. Jones," the nurse answered. "Where do you want it?" "Well, then," he answered with a smile, "I want it in your left arm, please."

02 이 일을 겪고 난 뒤 간호사가 느끼는 심정으로 가장 적절한 것은?

① 감사하다. ② 황당하다.
③ 답답하다. ④ 흐뭇하다.
⑤ 안타깝다.

03 밑줄 친 "I want it in your left arm, please."를 통해 남자가 말하고자 하는 의미는?

① 주사를 맞기가 아주 싫다.
② 치료를 제대로 해 달다.
③ 왼쪽 팔의 통증이 심하다.
④ 주사를 안 아프게 놓아 달라.
⑤ 간호사가 주사를 맞아야 한다.

UNIT 5 동사+목적어+목적보어

Prestudy

▣ 다음을 해석하시오.

1. She made her daughter a teacher.

____

2. The book makes me sleepy.

____

3. She told me to drive the car.

____

▣ 다음을 영작하시오.

1. 그의 아버지는 그를 의사로 만들었다.

____

2. 너의 미소는 나를 행복하게 해.

____

3. 그녀는 내가 유학 가기를 원한다.

____

Key Point

1. **5형식 문장 :** [주어+동사+목적어+보어]의 형태로 이루어진 문장을 말한다. 목적보어에는 형용사, 명사, 부정사, 원형부정사, 현재분사, 과거분사 등이 사용될 수 있다.
   She always makes me happy. [그녀는 항상 나를 행복하게 한다.]
   * 주어인 She, 동사인 makes, 목적어 me, 목적보어인 happy로 이루어진 문장이다.

2. 의문사+to 부정사의 종류

분류	형태	example
choose, elect, appoint, call, make 등	S+V+목적어+명사	They elected him mayor of the city. [그들은 그를 그 도시의 시장으로 선출했다.]
ask, advise, tell, get, expect, want, allow, order, cause 등	S+V+목적어+to부정사	She told me to do the dishes. [그녀는 내게 설거지를 하라고 말했다.] He advised me to exercise regularly. [그는 내게 규칙적으로 운동을 하라고 조언했다.]
make, have, let	S+V+목적어+원형부정사	He had the mechanic repair his car. [그는 정비공에게 그의 차를 수리 시켰다.]
see, watch, feel, hear 등	S+V+목적어+원형부정사	I saw her cross the river. [나는 그녀가 강을 건너는 것을 보았다.]
keep, make 등	S+V+목적어+형용사	He made her happy. [그는 그녀를 행복하게 만들었다.]

   지각동사의 경우 목적보어로 현재분사를 사용하게 되면 진행의 의미가 강조되고, 과거분사가 오게 되면 수동의 의미가 된다.
   I saw her crossing the street. [나는 그녀가 길을 건너고 있는 것을 보았다.]
   I heard my name called. [나는 내 이름이 불리는 것을 들었다.]

3. **5형식 문장의 구별**
   주어+동사+N1+N2에서 N1=N2의 관계가 성립한다.
   4형식 문장 : He gave me a book. [그는 나에게 책을 주었다.] (N1≠N2)
   5형식 문장 : We call him John. [우리는 그를 John이라고 부른다.] (N1=N2)

Basic Test

※ (　) 안에서 알맞은 말을 고르시오.

01 My parents didn't allow me (drive, to drive) a car.

02 I heard the lady (laugh, to laugh).

03 Mr. Smith asked her (play, to play) the piano.

04 Keep the classroom (clean, cleanly).

※ 다음 밑줄 친 부분을 어법에 맞게 고치시오.

05 My mother told me go to church every Sunday.

06 Tom made his friend to do his homework.

07 They wanted me speaking fast.

08 The new invention made him richly.

※ 우리말과 뜻이 같도록 주어진 단어를 배열하시오.

09 그는 내가 열심히 일하기를 원했다. (to work hard, me, wanted)

He ______________________.

10 그녀의 아기는 그녀를 행복하게 만든다. (makes, happy, her)

Her baby ______________________.

11 너는 우리 차가 움직이는 것을 느꼈니? (our car, move, feel)

Did you ______________________?

※ 밑줄 친 부분에 유의하여 우리말로 옮기시오.

12 I want them to play baseball with me.

____

13 The news made me sad.

____

Practice Test

※ 다음 빈칸에 들어갈 알맞은 말을 고르시오.

01 You'd better keep them _____ , or I will kick them out.

① silence ② silently
③ quiet ④ quietly
⑤ to quietly

02 The movie made her a great _________.

① acting ② actor ③ actress
④ action ⑤ active

03 Sally let us ______________ to the mall.

① go ② went ③ to go
④ going ⑤ goes

04 She wants __________________ Brian.

① marry ② married
③ marries ④ marrying
⑤ to marry

05 다음 중 어법상 틀린 것을 고르시오.

① She couldn't make her mother sad.
② I helped my mother carrying some foods.
③ He wanted us to join the music club.
④ She had her brother fix the computer.
⑤ He ordered us to stay in the classroom.

06 다음 중 어법상 틀린 것을 고르시오.

① She didn't ask him to bring the boxes.
② John had me take the little girl to the air port.
③ She saw me getting out of the building.
④ Terry heard him to speak English.
⑤ He told me to talk to you.

07 다음 우리말을 영어로 바르게 옮긴 것을 고르시오.

엄마는 내게 집에 있으라고 말씀하셨다

① Mom told me stay at home.
② Mom told me staying at home.
③ Mom told me to stay at home.
④ Mom told I stay at home.
⑤ Mom told to stay me at home.

08 다음 우리말을 영어로 바르게 옮긴 것을 고르시오.

Terry는 모두가 집에 일찍 가도록 했다.

① Terry lets everyone goes home early.
② Terry let everyone going home early.
③ Terry lets everyone to go home early.
④ Terry let everyone go home early.
⑤ Terry let everyone gone home early.

Reading Comprehension

01 다음 글의 밑줄 친 부분에서 알 수 있는 것은?

> Mrs. Mick wanted her husband to invite their neighbors, the Mack family, for supper. So Mr. Mick phoned up Mr. Mack. " And tomorrow?" " Nothing, either!" "What about the day after?" "We haven't planned anything for then, either!" "And Saturday?" "Oh, we are invited out then." "<u>That's too bad! We wanted to invite you to our house on Saturday!</u>"

① Mack 씨는 Mick 씨를 좋아하지 않는다.
② Mick 씨는 Mack 씨 집에 초대받고 싶어 한다.
③ Mick 씨는 Mack 씨의 말을 잘못 알아들었다.
④ Mick 씨는 Mack 씨 가족을 초대하고 싶어하지 않는다.
⑤ Mick 씨는 자신의 안타까움을 Mack 씨에게 전달하려고 한다.

※ 다음 글을 읽고 물음에 답하시오.

> One sunny day, four middle school boys didn't want to attend classes. So they were absent from school. The next morning they explained to a teacher why they missed her class. It is because their bus had a flat tire. Luckily she smiled and said, " Well, you missed a test yesterday, so take your seats and get out a pencil and paper." She waited while they sat down and got ready for the test. Then She said, "First question : Which tire was flat?"

02 이 일을 겪고 난 뒤 "they"가 느끼는 심정으로 가장 적절한 것은?

① 후련하다. ② 억울하다.
③ 안타깝다. ④ 부끄럽다.
⑤ 뿌듯하다.

03 이 글이 시사하는 바로 가장 적절한 것은?

① 쥐구멍에도 볕들 날 있다.
② 누구나 실수를 할 수 있다.
③ 진실은 드러나기 마련이다.
④ 뭉치면 살고 흩어지면 죽는다.
⑤ 백짓장도 맞들면 낫다.

대명사

UNIT 1 부정대명사 one, some, any

Prestudy

▣ 다음을 해석하시오.

1. I don't have a pen. Can I borrow one?

____

2. Come to the party if you have any free time.

____

▣ 다음을 영작하시오.

1. 너는 애완동물을 갖고 있니? -응, 하나 가지고 있어.

____

2. 탁자 위에 책이 몇 권 있다.

____

Key Point

1. **부정대명사 :** 범위가 정해지지 않은 막연한 사람이나 사물을 가리키는 말

1) one(단수개념인 것을 막연히 대신 받거나 가리킬 때 사용)

(1) 일반적인 사람들을 나타낼 때

One should keep one's promise

(2) 앞에서 언급된 불특정한 사물을 나타낼 때

A : Is there a PC room near here?

B : Yes, there's one across the street.

(3) 앞에 나온 명사를 받을 때

I haven't a pen. Can you lend me one?

* He took a stone and threw it.(it는 똑 같은 종류의 똑같은 바로 그 물건을 가리킬 때 사용, 주로 'the+명사'는 "it"로 받고, 'a+명사'는 "one"으로 받는다.)

2) some, any (어떤, 약간, 얼마간 대명사 또는 형용사로 쓰임)

(1) some : 긍정의 평서문, 권유의 뜻을 나타내는 의문문에서 사용

There is some milk on the table.

Will you have some coffee?

(2) any : 부정문, 의문문, 조건문에서 사용

Do you have any money?-No, I don't have any (money).

I don't know any foreign language.

3) 그 밖에 부정대명사

(1) one/the other : (둘 중) 하나는 ~이고, 나머지 하나는 ~이다.

Here are two pencils. (   ) is long, and (   ) is short.

[여기 연필 두 자루가 있다. 하나는 길고 하나는 짧다.]

(2) one/the others : (둘 이상일 때) 하나는 ~이고, 나머지들은 ~이다.

Here are five pens. (   ) is red, and (   ) are black.

[여기 다섯 개의 펜이 있다. 하나는 빨간색이고 나머지들은 검은색이다.]

(3) some/others : (정해지지 않은 다수 중에서) 일부는~, 다른 일부는~

Some people like soccer, but others like baseball.

[어떤 이들은 축구를 좋아하지만, 다른 이들은 야구를 좋아한다.]

(4) some/the others : (정해진 범위 안에서) 일부는~, 나머지 전부는~

Some of my classmate went to the movies, but the others didn't.

[반 친구 중 몇 명은 극장에 갔고, 나머지 일부는 가지 않았다.]

Basic Test

※ ( ) 안에서 알맞은 말을 고르시오.

01 I don't have a pen. Can you lend me (one, it)?

02 Alice found a nice skirt at the department store. She'll buy (it, one) tomorrow.

03 A : Is there any supermarket near here?
B : Yes, there is (it, one) on Silver Street.

04 He was trying to take a taxi, but he couldn't take (it, one).

05 Some people like cats and (other/others) don't.

06 He bought two balloons. One is blue and (another/the other) is red.

07 Here are five books. (One/Another) is for you, the others are for your sister.

08 Some of them went to library and (other/the others) didn't.

※ 다음 밑줄 친 부분을 바르게 고치시오.

09 A : Is this blue shirt yours?
B : No, one isn't.

10 Water is very useful. We must not waste them.

11 A : We have many books.
B : Can you give me a good ones?

12 The baby has three balls; a large one, and two small one.

※ 밑줄 친 부분이 의미하는 것을 설명하려고 할 때, 빈칸에 알맞은 단어를 쓰시오.

13 A : What is a panda like?
B : It is like a bear. 〈It=a ____________〉

14 I don't have a notebook computer. I want to buy one. 〈one=a ________________〉

※ 다음에서 어법상 틀린 부분을 찾아 바르게 고치시오.

15 A : Do you need any color pencils?
B : Yes, I need some red one.

16 I bought a car yesterday. One is very expensive.

17 I don't have some money.

Practice Test

01 다음 빈칸에 들어갈 알맞은 말을 고르시오.

> I need a pen. Do you have ______?

① it ② that
③ those ④ one
⑤ this

02 다음 빈칸에 들어갈 알맞은 말을 고르시오.

> A : Will you drink __________ coffee?
> B : Yes, thanks.

① one ② no
③ any ④ those
⑤ some

03 다음 빈칸에 들어갈 알맞은 말을 고르시오.

> A : Where is my cell phone?
> B : ______ is on the sofa. Don't sit on it.

① One ② There
③ It ④ This
⑤ These

04 다음 빈칸에 들어갈 알맞은 말을 고르시오.

> There aren't __________ empty seats.

① one ② no
③ any ④ those
⑤ some

05 다음 중 어법상 틀린 것을 고르시오.

① One should follow the rules.
② Do you have any questions?
③ Will you have some bread?
④ I don't have some cats.
⑤ I want some music files.

06 다음 중 어법상 틀린 것을 고르시오.

① I'll give her some comic books.
② Did you see any dogs near here?
③ Don't you need any erasers?
④ I bought some new clothes.
⑤ There is any milk in the box.

07 밑줄 친 우리말을 영어로 바르게 옮긴 것은?

> I have two dogs. 한 마리는 검은색이고, 다른 한 마리는 갈색이다.

① It is black and brown.
② It is black, the other it is brown.
③ One is black, the other is brown.
④ One is black, the second is brown.
⑤ One is black, others are brown.

Reading Comprehension

01 다음 글의 큰 아들을 묘사하는 말로 가장 알맞은 것은?

A father said to his sons : "Tomorrow your mother is going to bake a pie. Who is going to eat it?" The oldest son replied : " Father, I'll eat it all!" The father then said : "Tomorrow I am going to buy a pig. Who is going to eat it?" The same son answered : "Father, I'll eat it all."
The father added : "Tomorrow I'm going to send one of you on an errand. Who is going to help me?"
The oldest son answered again : "It has always been me. Now it's somebody else's turn to help."

① 헌신적이다. ② 반항적이다.
③ 부지런하다. ④ 약삭빠르다.
⑤ 수줍음이 많다.

02 다음 글의 내용상 빈 칸에 가장 알맞은 말은?

The human body is a whole. When one part of the body is ill, the other parts feel ill, too. For example, when you are hit with a cold or have the flu, your nose runs and you have a cough and a headache. You feel weak and tired. Your mouth becomes dry, so your food doesn't taste right and you don't want to eat anything. Your brain works slowly than usual, and nothing seems fun. A cold is not usually a serious illness, but it shows us ______________.

① why human body functions in parts
② what we should do when we catch a cold
③ why the brain slows down when we're sick
④ how important it is to take care of our body
⑤ how the various parts of our body are connected

UNIT 2 재귀대명사

Prestudy

▣ 다음을 해석하시오.

1. He made a promise to himself.

____

2. Did you make the bread yourself?

____

▣ 다음을 영작하시오.

1. 제 소개를 해도 될까요?

____

2. 그는 혼잣말을 하고 있다.

____

3. 내가 직접 할 거야.

____

Key Point

1. **재귀대명사** : 소유격 또는 목적격에 -self/-selves가 붙어 '~자신'이란 뜻이 되는 대명사, 동사나 전치사의 목적어로 사용되어 주어의 동작이 주어 자신을 나타낸다.

	단수	복수
1인칭	I→myself	we→ourselves
2인칭	you→yourself	you→yourselves
3인칭	she/he/it→herself/himself/itself	they→themselves

1) 재귀 용법 : 주어가 하는 동작이 주어 자신에게 되돌아올 때, 즉 주어와 목적어가 같은 사람일 때 동사의 목적어나 전치사의 목적어로 사용되는 재귀대명사를 재귀용법이라고 한다. 이 때 재귀대명사는 생략할 수 없다.

I looked at myself in the mirror. (I=myself)

Tom and Ann blamed themselves for the accident. (Tom and Ann=themselves)

2) 강조용법 : 주어, 목적어, 보어의 뜻을 강조하기 위해 쓰인 재귀대명사를 강조용법이라고 하며, "~자신이 직접" 정도의 의미를 가진다. 강조용법으로 쓰인 재귀대명사는 문장에 반드시 필요한 문장성분이 아니므로 생략할 수 있다. 재귀대명사는 강조하는 말 바로 뒤에 오지만, 주어를 강조하는 경우 문장의 뒤에 올 수도 있다.

Andy himself wrote the story.

I made this home page myself.

3) 전치사+재귀대명사

(1) for oneself=for one's sake : 자기 자신을 위하여
=without another's help : 혼자 힘으로

(2) by oneself=all alone : 외로이, 홀로

He lives in that house by himself.

(3) 기타

① The candle went out of itself. : 저절로

② to oneself : 자기 혼자에게만

③ beside oneself=insane=mad=at the end of self-control : 제정신이 아닌, 미친

④ come to oneself=become conscious : 의식을 찾다, 제정신이 들다.

⑤ pride oneself on=be proud of=take (a) pride in : 자랑하다.

⑥ absent oneself from=be absent from : 출석하지 않다.

⑦ present oneself at=be present at : 출석하다.

⑧ avail oneself of : ~을 이용하다.

⑨ accustom oneself to : ~에 익숙해지다.

⑩ overdrink oneself : 과음하다.

⑪ overwork oneself : 과로하다.

⑫ oversleep oneself : 늦잠자다.

⑬ in itself : 본질적으로

Basic Test

※ (　) 안에서 알맞은 말을 고르시오.

01 She often draws a picture of (himself, herself).

02 Did you enjoy (you, yourself)?

03 We must do our homework (ourself, our selves).

04 A : This soup is delicious.
B : I cooked it (me, myself).

※ 다음 밑줄 친 부분을 어법에 맞게 고치시오.

05 We enjoyed <u>us</u> last night.

06 He was very proud of <u>herself</u>.

07 I wrote this book <u>me</u>.

※ 우리말과 뜻이 같도록 빈칸에 알맞은 말을 쓰시오.

08 과일을 마음껏 드세요.
Please help ________ to the fruit.

09 그는 숙제를 스스로 했다.
He did his homework ________________.

10 역사는 되풀이된다.
History repeats ____________________.

※ 밑줄 친 부분에 유의하여 우리말로 옮기시오.

11 He should clean his room <u>himself</u>.

____

12 Tim looked at <u>himself</u> in the mirror.

____

13 Sometimes I talk to <u>myself</u>.

____

Practice Test

※ 다음 빈칸에 들어갈 알맞은 말을 고르시오.

01 Welcome. Please introduce ______ , Mrs. Jang.

① yourself ② himself ③ herself
④ you ⑤ myself

02 There's some bread on the table. Just help ______ to it.

① oneself ② itself ③ yourself
④ myself ⑤ herself

03 She likes to talk about ______________.

① himself ② herself ③ yourself
④ themselves ⑤ itself

04 다음 중 밑줄 친 부분을 생략할 수 없는 것은?

① He killed himself 13 years ago.
② You don't have to throw it away yourself.
③ She finished it herself.
④ She herself did the work.
⑤ They fixed the MP3 player themselves.

05 다음 중 어법상 틀린 것을 고르시오.

① She herself will take care of him.
② He was too proud of himself sometimes.
③ She went to see me yourself.
④ They couldn't enjoy themselves at all.
⑤ Did you write it down yourself?

06 다음 중 어법상 틀린 것을 고르시오.

① Amy likes to take photos of herself.
② Jane was almost beside herself.
③ I made myself some gimbap.
④ I wrote her a letter by herself.
⑤ Did you make this cake by yourself?

07 다음 우리말을 영어로 바르게 옮긴 것을 고르시오.

그녀는 모든 것을 혼자 힘으로 해냈다.

① She did everything on herself.
② She did everything for her.
③ She did everything for itself.
④ She did everything by oneself.
⑤ She did everything by herself.

08 다음 우리말을 영어로 바르게 옮긴 것을 고르시오.

너 다쳤니?

① Did you hurt yourself?
② Did yourself hurt you?
③ Did you hurt myself?
④ Did you hurt you?
⑤ Did you hurt oneself?

Reading Comprehension 

※ 다음 글을 읽고 물음에 답하시오.

A long time ago, two friends lived together in a small cottage. Their names were John and Bobby. John and Bobby were not rich, but they were happy. They were never lonely because they had each other. They had a simple but good life. Then, in the spring of 1858, John became ill and died. After John was buried, Bobby stayed near John's grave. For 14 years Bobby never left there. Finally, in 1872, Bobby died, too. Friends buried him in a little grave beside John. Bobby, John's best friend, was a dog.

01 Bobby를 설명하는 말로 가장 적절한 것은?

① 지혜롭다. ② 용감하다.
③ 활발하다. ④ 충성스럽다.
⑤ 잘 생겼다.

02 Bobby에 관한 설명 중 글의 내용과 일치하지 않는 것은?

① 오두막집에서 주인과 같이 살았다.
② 여러 이웃과 사이좋게 지냈다.
③ 한시도 주인 곁을 떠나지 않았다.
④ 주인이 죽은 뒤에도 떠나지 않았다.
⑤ 죽어서도 주인 무덤 곁에 묻혔다.

※ 다음 글을 읽고 물음에 답하시오.

You have failed many times, although you may not remember. You fell down when you tried to walk for the first time. You almost drowned when you tried to swim for the first time, didn't you? Did you hit the ball when you first swung a bat? Even Babe Ruth struck out 1,330 times, but he also hit 714 home runs. Don't worry about failure. You will miss many chances if you don't even try. Worry about them.

03 밑줄 친 them 이 가리키는 것으로 가장 적절한 것은?

① balls ② failures
③ chances ④ home runs
⑤ bats

04 필자가 주장하는 바로 가장 적절한 것은?

① 실패를 두려워 하지 말자.
② 남은 인생에 최선을 다하라.
③ 고난을 인내함으로 이겨내라.
④ 시간을 낭비하지 말고 아껴라.
⑤ 시간은 아무도 기다려 주지 않는다.

형용사와 부사

- UNIT 1. 형용사와 부사의 역할
- UNIT 2. 비교 표현(원급, 비교급, 최상급)

UNIT 1 형용사와 부사의 역할

Prestudy

▣ 다음을 해석하시오.

1. You have a nice house.

____

2. The 63 Building is very tall.

____

3. Tom ran to the house very fast.

____

4. My teacher is always kind to me.

____

▣ 다음을 영작하시오.

1. 우리는 커다란 집에서 산다.

____

2. 나의 여동생은 매우 예쁘다.

____

3. 나는 종종 도서관에 간다.

____

Key Point

1. **형용사 :** 명사나 대명사를 수식하거나 설명해 주는 역할

1) 수식 : 명사의 앞이나 뒤에서 꾸며준다.
   She has long→hair.

※ −thing, −body, −one으로 끝나는 대명사는 반드시 뒤에서 형용사의 수식을 받는다.
   something←cold(○), cold something(×)

2) 설명 : 주어나 목적어의 보어로 사용된다.
   Jack is smart and strong.(주격 보어)
   I kept the door open.(목적격 보어)

2. **부사 :** 동사, 형용사, 다른 부사, 문장 전체를 수식하는 역할
1) 부사의 종류 : 뜻에 따라 5가지로 분류할 수 있습니다.
   ① 때는 나타내는 부사 : soon(곧), yet(아직), then(그때), now(지금), today(오늘), yesterday(어제), early(일찍)
   ② 장소를 나타내는 부사 : here(여기), there(저기), far(멀리), near(가까이), up(위로), down(아래로)
   ③ 방법을 나타내는 부사 : well(잘), loudly(크게), carefully(조심스럽게), quickly(빨리)
   ④ 빈도를 나타내는 부사 (얼마나 자주 발생하는 지를 나타내는 단어)

100%	80~90%	60~70%	30~40%	10%	0%
always	usually	often	sometimes	seldom	never
항상	대개	자주	가끔	거의~않다.	절대~않다

⑤ 정도를 나타내는 부사 : so (너무), too (너무), very (매우), quite (꽤)

2) 부사의 역할
   ① 동사 수식 : Tom drives carefully.

② 형용사 수식 : This pizza tastes really good.

③ 부사수식 : Billy speaks English quite well.

④ 문장 전체 수식 : Luckily, we had some money.

3) 부사의 형태

형용사+ly : 형용사 뒤에 ly를 붙여서 부사를 만들 수 있습니다.

① really, carefully, beautifully, sadly

② happily, easily, luckily, busily→y로 끝난 형용사는 y를 i로 바꾸고 ly를 붙여줍니다.

③ simply, gently, possibly, terribly→e로 끝난 형용사는 e를 빼고 ly를 붙여줍니다.

※ 형용사/부사의 형태가 같은 단어

early : 이른/일찍, fast : 빠른/빨리, late : 늦은/늦게, hard : 어려운/열심히, high : 높은/높이

Basic Test

※ 다음 밑줄 친 부분을 어법에 맞게 고치시오.

01 He made me <u>happily</u>.

02 Sharon looks <u>kindly</u>.

03 I would like to drink <u>cold something</u>.

04 He works <u>diligent</u>.

※ 우리 말과 뜻이 같도록 (　) 안에서 알맞은 말을 고르시오.

05 그는 매우 영리한 아내가 있다.
He has a very (smart, smartly) wife.

06 분명하게 말하시오.
Speak (clear, clearly).

07 그는 종종 컴퓨터 게임을 한다.
He (plays often, often plays) computer games.

※ 우리말과 뜻이 같도록 빈칸에 알맞은 말을 쓰시오.

08 너무 빨리 걷지 마세요.
Please don't walk so ____________________.

09 그는 집으로 천천히 걸어갔다.
He walked home ____________________.

10 나는 그 책이 재미있다는 것을 알았다.
I found the book ____________________.

※ 우리말과 뜻이 같도록 주어진 단어를 배열하시오.

11 제 말을 주의 깊게 들으세요. (carefully, listen, me, to)
____________________.

12 그녀는 항상 나에게 미소 짓는다. (smiles, always, me, at)
She ____________________.

13 오늘은 날씨가 화창해요. (today, lovely, is)
It ____________________.

Practice Test

※ 다음 빈칸에 들어갈 알맞은 말을 고르시오.

01 He can run ______________.

① fastness ② with fastly
③ fast ④ fasting
⑤ fastly

02 I want ______________ to eat.

① cold something
② some cold
③ cold some
④ something cold
⑤ cold a thing

03 다음 빈칸에 들어갈 수 없는 말은?

I don't think his sister looks ______________.

① ugly ② pretty ③ beautiful
④ friendly ⑤ heavily

04 다음 빈칸에 들어갈 말이 순서대로 바르게 짝지어진 것은?

His mother made him work ______________.
He always listened to her advice ______________.

① hardly−careful
② hard−carefully
③ hard−carelessly
④ hardly−carefully
⑤ hard−careful

05 다음 중 어법상 틀린 것을 고르시오.

① I kept the window open.
② You should read the book carefully.
③ That pizza smells deliciously.
④ He wants something hot to drink.
⑤ Subin has a pretty sister.

06 다음 중 어법상 틀린 것을 고르시오.

① John often cooks for his mother.
② Tony usually gets up at about 7.
③ Lance sometimes is late for work.
④ Clara will always love you.
⑤ In fact, Owen never lied to us.

07 다음 우리말을 영어로 바르게 옮긴 것을 고르시오.

그녀는 항상 화장을 짙게 한다.

① She always wear heavy makeup.
② She wears heavy always makeup.
③ She wears always heavy makeup.
④ She always wears heavy makeup.
⑤ She always heavy wears makeup.

Reading Comprehension

※ 다음 글을 읽고 물음에 답하시오.

Cats and dogs are both popular pets. But cats are better pets in some ways. First, cats are cleaner. They love to be very clean. Cats are also quieter than dogs. They usually do not make a lot of noise. Cats are safer, too. Dogs sometimes bite people, but cats almost never do. You can easily look after cats. And you do not need to spend much time with a cat. In fact, many cats like to be ______________.

01 빈 칸에 들어갈 알맞은 말은?

① alone ② clean
③ active ④ noisy
⑤ easy

02 고양이가 개보다 더 나은 이유로 적절하지 않은 것은?

① 깨끗하기 때문에
② 조용하기 때문에
③ 안전하기 때문에
④ 작고 귀엽기 때문에
⑤ 돌보기가 쉽기 때문에

※ 다음 글을 읽고 물음에 답하시오.

People drink water every day. A camel does not. A camel can go without water for a long time. Some people think that a camel's hump is filled with water. That is not true. A camel's hump is filled with fat. a camel uses the fat in its hump for food. So a camel can go without ______________ for a long time, too. What a strange animal the camel is!

03 밑줄 친 That 이 나타내는 말은?

① a camel's hump is filled with fat
② a camel's hump is filled with water
③ a camel uses the fat in its hump for food
④ a camel can go without water for a long time
⑤ a camel is a very strange animal

04 빈 칸에 들어갈 알맞은 말은?

① resting ② eating
③ drinking ④ sleeping
⑤ walking

UNIT 2 비교 표현(원급, 비교급, 최상급)

Prestudy

▣ 다음을 해석하시오.

1. Tony is as handsome as his father.

____

2. KTX is faster than the subway.

____

3. I think that he's the best singer in Korea.

____

▣ 다음을 영작하시오.

1. 그는 Brian만큼 키가 크다.

____

2. 그녀는 20살보다 더 나이 들어 보인다.

____

3. 이것은 이 거리에서 가장 오래된 건물이다.

____

Key Point

1. **원급〈as+형용사/부사의 원급+as〉** : ~만큼~하다.
   Today is as cold as yesterday.

2. **비교급〈비교급(−er/more)+than…〉** : ~보다 더 ~한
   You are better than me.
   Health is more important than money.
   * even, much, a lot, still, far 등은 비교급 앞에 쓰이면 '훨씬'의 의미로 비교급 강조.

3. **최상급 the+최상급(−est/most)** : 가장~한
   I'm the oldest son in my family.
   He is one of the most famous musician in Korea.
   * one of the 최상급+복수명사 : 가장… 한 ~중의 하나

4. **비교급과 최상급을 만드는 방법**

1) 규칙변화 형용사/부사

	이론	원급	비교급	최상급
1	형용사와 부사의 원급에 'er, est' 를 붙인다.	fast	faster	fastest
		small	smaller	smallest
2	원급의 어미가 'e'로 끝나면 '~r, ~st' 를 붙인다.	nice	nicer	nicest
		large	larger	largest
3	'단모음+단자음'으로 끝나면 자음 하나 더 쓰고 'r, st'를 붙인다.	big	bigger	biggest
		hot	hotter	hottest
4	'자음+y'로 끝나면 'y'를 'i'로 바꾸고 'er, est'를 붙인다.	happy	happier	happiest
		easy	easier	easiest
5	2음절 대부분과 3음절 이상의 단어에는 'more, most'를 붙인다.	famous	more famous	most famous
		beautiful	more beautiful	most beautiful

2) 불규칙 변화 형용사/부사

	원급	비교급	최상급	의미
1	good/well	better	best	좋은/잘
2	bad/badly	worse	worst	나쁜/아픈
3	many/much	more	most	많은(수)/많은(양)
4	little	less	least	양이 적은

Basic Test

※ 우리말과 뜻이 같도록 주어진 단어를 알맞은 형태로 바꾸어 쓰시오.

01 금은 구리보다 무겁다. (heavy)

Gold is __________ than copper.

02 그는 전보다 더 행복하다. (happy)

He is ____________ than before.

03 이 문제는 그것보다 쉽다. (easy)

This question is ____ than that.

04 이 TV 프로그램이 저것보다 더 흥미롭다. (exciting)

This TV program is ______ than that one.

※ 다음 밑줄 친 부분을 바르게 고치시오.

05 He is <u>tall</u> than his father.

06 It was <u>bad</u> than I thought.

07 He is as <u>smarter</u> as his father.

※ (   ) 안에서 알맞은 말을 고르시오.

08 This flower is (prettier, more pretty) than that one.

09 Ted runs (faster, fastest) than Jim.

10 Your house is (large, larger) than mine.

※ 우리말과 뜻이 같도록 주어진 단어를 배열하시오.

11 그는 나보다 나이가 많다. (I, than, older)

He is ____________________________.

12 Mike는 우리 반에서 가장 수영을 잘한다. (swimmer, best, the)

Mike is ________________ in our class.

13 나는 올해 가장 흥미로운 영화를 보았다. (most, the, exciting)

I watched ____________ movie this year.

Practice Test

※ 다음 빈칸에 들어갈 알맞은 말을 고르시오.

01 Yujin is as ________________ as Taehi.

① smaller ② small
③ the smallest ④ more small
⑤ most small

02 Your health is ______ than your money.

① important ② importanter
③ importantest ④ more important
⑤ most important

03 He may be __________ boy in his class.

① heavier ② more heavier
③ the heaviest ④ more heavy
⑤ the most heavy

04 다음 중 어법상 <u>틀린</u> 것을 고르시오.

① He's the most popular singer in Korea.
② She can run faster than Karl.
③ Jiseop became famouser than Inseong.
④ He is as handsome as Wooseong.
⑤ She is more intelligent than him.

05 다음 중 어법상 <u>틀린</u> 것을 고르시오.

① My cell phone looks better than yours.
② Who is the strongest man in the world?
③ Your room is much cleaner than his.
④ This jacket is most expensive than the black one.
⑤ You look as pretty as her.

06 다음 우리말을 영어로 바르게 옮긴 것을 고르시오.

날씨가 점점 추워진다.

① It is getting cold and cold.
② It is getting more colder and colder.
③ It is getting cold and colder.
④ It is getting colder and colder.
⑤ It is getting colder but colder.

07 다음 우리말을 영어로 바르게 옮긴 것을 고르시오.

그녀는 그만큼 친절하지 않아.

① She isn't as kind as him.
② She is as not kind as him.
③ She is so not kind as him.
④ She is as kind as him.
⑤ She isn't as him as kind.

Reading Comprehension

01 다음 글의 주제문은?

(A) But Leonardo wasn't just an inventor. (B) He was one of the greatest artists of his days. (C) By the time he was twenty years old, he was called a master painter, and as he got older, he became even more famous. (D) Like his scientific inventions, (E) He was always trying a new paint or some new colors.

① A ② B ③ C
④ D ⑤ E

※ 다음 글을 읽고 물음에 답하시오.

One wet day I was on my way in the bus. Then a woman with a dog got on the bus. It was a big dog. It was about as big as a young house. Its feet were very dirty. The woman said, "If I pay for my dog, can he have a seat like other passengers?" The bus driver looked at the dog and then he said, "Certainly, madam. He can have a seat like all the other passengers. But, like other passengers, he must not put his feet on it."

02 From this story we can say the bus driver did not want ______________ to get dirty.

① the seat ② the woman
③ the dog's feet ④ himself
⑤ passengers

03 Which was not true in this story?

① A woman with a dog got on the bus.
② "I" was riding beside the dog.
③ The dog was very big.
④ The dog's feet were very dirty.
⑤ She wanted her dog to sit on a seat.

접속사

UNIT 1 등위접속사 and, but, or

Prestudy

▣ 다음을 해석하시오.

1. He plays the guitar and sings.

____

2. He can speak English, but he cannot speak Chinese.

____

3. Do you go to school by bus or by subway?

____

▣ 다음을 영작하시오.

1. 북극곰은 키가 크고 하얗다.

____

2. 제인은 과학을 좋아하지만, 나는 수학을 좋아한다.

____

3. 차가운 차 마실래요 뜨거운 차 마실래요?

____

Key Point

1. **등위 접속사** : 문법적으로 대등한 역할을 하는 단어-단어, 구-구, 절(문장)-절(문장)을 이어 주는 역할을 하며, and, but, or 등이 있다.

(1) and : 그리고, 그래서

I'll buy some milk and bread.

Why don't you sit down and rest for a while?

(2) but : 그러나, 그런데

Tom is American, but he likes kimchi.

I can remember her face, but I can't remember her name.

(3) or : 혹은, 또는

Is today Wednesday or Thursday?

Which do you like better, summer or winter?

2. **상관 접속사**

(1) both(at once, alike) A and B : A도 B도 ~이다→양자긍정, 복수동사

He is both(at once, alike) tired and hungry.

[그는 지치기도 하고 배고프기도 하다.]

He is both tired and hunger.(×)〈형용사 and 명사〉

* 요즈음은 A and B alike 의 형태로 쓴다.

(2) neither A nor B : A도 B도 ~아니다→양자부정, 근자일치

I like neither tea nor coffee. [나는 홍차도 좋아하지 않고 커피도 좋아하지 않는다.]

=I don't like either tea or coffee.

(3) either A or B : A나 B 둘 중에 하나→양자택일, 근자일치

Either you or I am in the wrong.

[너든 나든 어느 한쪽이 잘못이다.]

He was either drunk or crazy.

[그는 술 취했거나 또는 미쳤거나 어느 한 쪽이었다.]

(4) not only(merely, simply, solely, alone, just) A but (also) B

=B as well as A : A뿐만 아니라 B도 역시 ~이다. →B>A, 강조일치

Not only you but also he is tall. =He as well as you is tall.

[당신뿐만 아니라 그 사람도 키가 크다.]

(5) not so much A as B=much more B than A=B, rather than A : A 라기보다는 B다.

He is not so much a scholar as a professor.

He is much more a professor than a scholar.

He is a professor, rather than a scholar.

[그는 학자라기보다는 교수다.]

Basic Test

※ ( ) 안에서 알맞은 말을 고르시오.

01 Yuseong did his homework (and, or) went out.

02 I am poor, (and, but) I am happy.

03 Which do you want more, money (but, or) love?

04 Jason can speak English, (but, or) he can't speak French.

※ 우리말과 뜻이 같도록 빈칸에 알맞은 말을 쓰시오.

05 영민이는 착하고 똑똑하다.
Yeongmin is nice ________________ smart.

06 이 핸드백은 비싸지만 유용하지는 않다.
This handbag is expensive _______ useless.

07 걸어서 갈거니, 버스로 갈거니?
Will you go on foot ___________ by bus?

※ 우리말과 뜻이 같도록 밑줄 친 부분을 알맞게 고치시오.

08 흥부는 춥고 배고팠다.
Heungbu was cold <u>but</u> hungry.

09 이번 여름에는 산으로 갈까, 바다로 갈까?
Where do you want to go this summer, to the mountains <u>and</u> the sea?

10 Ted는 키는 크지만 농구는 잘 못한다.
Ted is tall <u>or</u> he cannot play basketball well.

※ 밑줄 친 부분에 유의하여 우리말로 옮기시오.

11 Minhee <u>and</u> I are good friends.

____

12 Let's go <u>and</u> help him.

____

13 Would you like some tea <u>or</u> coffee?

____

Practice Test

※ 다음 빈칸에 들어갈 알맞은 말을 고르시오.

01 She is beautiful, ________ isn't smart.

① if ② but
③ because ④ after
⑤ that

02 Which do you like better, milk ________ juice?

① and ② that
③ or ④ but
⑤ so

03 Jina ______ Jihyeon are always together.

① and ② but
③ or ④ so
⑤ because

04 다음 빈칸에 들어갈 말이 순서대로 짝지어진 것은?

• She went to the PC room, ________ she's at home now.
• Is today Saturday ________ Sunday?

① and−and ② or−but
③ but−or ④ so−or
⑤ because−and

05 다음 중 어법상 틀린 것을 고르시오.

① Jim and Tim are twins.
② I like to watch movies, but I don't like action movies.
③ She likes me or he doesn't like me.
④ I'll have a sandwich and an apple.
⑤ I ran in the marathon, but I didn't win.

06 다음 중 어법상 틀린 것을 고르시오.

① Jin is wise and diligent.
② Maria is American, but she can speak Korean.
③ I ran hard, but I missed the bus.
④ Henry practiced tennis a lot, but he's good at it now.
⑤ Daniel is handsome, but he's not popu lar with girls.

07 다음 우리말을 영어로 바르게 옮긴 것은?

그는 일본인이 아니라 한국인이다.

① He is not Japanese, but Korean.
② He is not Japanese, and Korean.
③ He is not Japanese, or Korean.
④ He is no Japanese, but Korean.
⑤ He is but Japanese, not Korean.

Reading Comprehension

01 다음 글은 chameleon(카멜레온)의 어떤 특성에 대한 설명인가?

> Have you ever seen a chameleon? Do you know what it is ? It is a slow-moving animal that is able to change color. One minute it may be yellow or green, and the next moment it may be dark grey. It can be protected from other animals by changing color like that.

① 명암을 구별하는 방법
② 먹이를 잡는 방법
③ 다른 장소로 이동하는 방법
④ 자신을 보호하는 방법
⑤ 색깔을 바꾸는 방법

※ 다음 글을 읽고 물음에 답하시오.

> At one time, there were no women doctors in the United States. Many people thought that women should not be doctors. But Elizabeth Blackwell did not agree. She wanted to be a doctor. She wanted to help people. Elizabeth Blackwell studied hard. In 1849, she became the first woman doctor in the United States. Dr. Blackwell spent her time caring for poor people. She also started a school to teach other women to be doctors.

02 "Elizabeth Blackwell"에 대한 설명 중 글의 내용과 일치하지 않는 것은?

① 가난한 사람을 도와 주고 싶었다.
② 의사가 되려고 최선을 다했다.
③ 미국 최초의 여의사가 되었다.
④ 여성무료 병원을 설립하였다.
⑤ 후진 양성에 힘을 기울였다.

03 밑줄 친 Elizabeth Blackwell did not agree.의 의미를 다음 문장으로 나타낼 때 빈칸에 적절한 것은?

> →She thought that she could ____________.

① study hard　　② be a doctor
③ start a school　　④ help poor people
⑤ teach other women.

New Words change : 바꾸다
protect : 보호하다

UNIT 2 종속접속사 when, before, after

Prestudy

▣ 다음을 해석하시오.

1. When I was in Europe, I made many friends.

____

2. I leaned my house before my friends visited me.

____

3. After he bought a new car, he drove every day.

____

▣ 다음을 영작하시오.

1. 나는 어렸을 때 개를 키웠다.(when)

____

2. 길을 건너기 전에 주의해라.(before)

____

3. 그들은 경기에 이긴 후, 파티를 열었다.(after)

____

Key Point

1\. **종속 접속사** : 주절과 종속절을 이어 주는 접속사로, 종속적은 주절의 내용을 보충하는 역할을 한다.

1) 시간을 나타내는 종속 접속사

(1) when : ~할 때

When I was young, I was pretty.

Please call me when you have time.

(2) before : ~하기 전에

I asked a question before the class ended.

Before you leave, turn off the light, please.

(3) after : ~한 후에

I remembered the answer after I turned in my exam.

After they succeeded, they helped poor people.

(4) as : '~할 때에', '(한창)~하고 있을 때', '~하면서 (when, while)'

She sang as she washed.

As the master came into the room, they stopped talking.

(5) since : '~이래 (줄 곧=from the time when)'

I have known him since he was a boy.

It is(has been) five years since he began writing novels.

(6) as soon as : '~하자마자,' '~하자 곧'

I told him so as soon as he got the phone call.

(7) while : ~동안에

He read a book while he waited for her.

While I'm having lunch, you can go swimming.

* while : ~인 반면에, ~이지만

While I agree with you, I don't believe that your way is best.

Basic Test

※ 다음 밑줄 친 부분에 유의하여 내용이 자연스럽게 이어지도록 연결하시오.

01 I saw a rainbow	•	• ⓐ he showed them to me.	
02 After he took some pictures,	•	• ⓑ you should learn English.	
03 Before you go to America,	•	• ⓒ after it stopped raining.	

※ 우리말과 뜻이 같도록 밑줄 친 부분을 알맞게 고치시오.

04 내가 그곳에 도착했을 때 비가 오고 있었다.

Before I got there, it was raining.

05 그는 시카고에 가면 힐튼호텔에 머물 것이다.

He will stay at Hilton Hotel before he goes to Chicago.

06 자기 전에 나는 TV를 볼 것이다.

When I go to bed, I will watch TV.

07 일을 마친 후 나는 휴식을 취할 것이다.

I'll take a rest before I finish the work.

※ 우리말과 뜻이 같도록 빈칸에 when, before, after 중 알맞은 접속사를 쓰시오.

08 선생님이 화내시기 전에 우리 조용히 하자!

Let's be quiet________ the teacher gets angry.

09 눈이 올 때 나는 행복하다.

________ it snows, I feel happy.

10 Ann은 항상 학교 가기 전에 아침을 먹는다.

Ann always eats breakfast__________ she goes to school.

11 오늘 밤 저녁을 먹은 후에 나는 숙제를 할 것이다.

I'm going to do my homework __________ I eat dinner tonight.

※ 밑줄 친 부분에 유의하여 우리말로 옮기시오.

12 When I grow up, I will be a lawyer.

____

13 I missed him after he left me.

____

Practice Test 

※ 다음 빈칸에 들어갈 알맞은 말을 고르시오.

01 ______ I was 12, I went to Canada to study.

① But ② After ③ Or
④ When ⑤ That

02 We have some time ______ the show starts.

① that ② but
③ or ④ before
⑤ because

03 You can go out ______ you wash the dishes.

① after ② then
③ or ④ that
⑤ for

04 다음 빈칸에 들어갈 말이 순서대로 짝지어진 것은?

- ______ you leave, turn off the TV.
- My family was poor ______ I was young.

① After−if ② When−before
③ Before−when ④ Before−but
⑤ So−before

05 다음 중 어법상 틀린 것은?

① We have to clean the room before she comes back.
② After they started to make a lot of money, everything changed.
③ Before she became a singer, she couldn't dance at all.
④ We're going to meet after finish I cooking.
⑤ Before she died, she helped many poor people.

06 다음 중 밑줄 친 부분의 쓰임이 나머지 넷과 다른 것은?

① Everyone was watching the news when I came home.
② When did you hear the news?
③ How did you feel when you met her for the first time?
④ Jennifer was crying when I went to see her.
⑤ When you see him, tell him to call me.

Reading Comprehension

01 다음 글에서 필자가 고민하고 있는 문제로 알맞은 것은?

Dear Darcy,

My boyfriend only loves himself. He only thinks about himself. He stops and looks at himself when he passes by a store window. He doesn't buy me a gift on my birthday, and he never brings me flowers. And that's not all! He's also very selfish with his things. For example, he never lets me use his cell phone. He says that it is his and he doesn't want me to use it. What should I do?

Unhappy

① 남자 친구가 과소비를 하는 것
② 남자 친구가 다른 여자를 만나는 것
③ 남자 친구가 연락을 자주 하지 않는 것
④ 남자 친구가 이기적인 모습을 보이는 것
⑤ 남자 친구가 외모에 너무 신경쓰는 것

02 다음 글의 내용을 한 문장으로 요약 할때 빈칸 (A)와 (B)에 알맞은 말로 바르게 짝지어진 것은?

American movies are full of fast cars and car chases. The bad guys drive through red traffic lights, and the police do the same. That's in the movies. Real life is very different from the movies. You will be surprised when you go to the U.S. People drive carefully, and the traffic is very slow. It is not a good idea to break the speed limit. The highway police out radar and cameras along the road. And they have helicopters! You can't see a police car in your mirror, but look up. The pilot above you may be reporting you to a Highway Patrol officer!

→In the ____ (A) ____ Americans drive fast. But in real life they drive ____ (B) ____ .

	(A)		(B)
①	reality	–	carelessly
②	reality	–	carefully
③	movies	–	carefully
④	movies	–	carelessly
⑤	highway	–	carelessly

UNIT 3 종속접속사 because, if, that

Prestudy

▣ 다음을 해석하시오.

1. Sally didn't go to work because she was sick.

____

2. What do we do if it rains?

____

3. She hopes that her son will succeed.

____

▣ 다음을 영작하시오.

1. 날씨가 추웠기 때문에 그녀는 창문을 닫았다.

____

2. 내가 복권에 당첨된다면, 나는 세계 여행을 할 것이다.

____

3. 나는 네가 그것을 할 수 있다고 생각해.

____

Key Point

1. 이유, 조건을 나타내는 종속 접속사

(1) because (원인, 이유) : ~이기 때문에

Because I was tired, I couldn't finish my homework.

(2) if (조건) : ~만약…하다면

If you love her, you'll want to know everything about her.

(3) since : ~때문에,~이므로

Since you look tired, you had better take a rest.

(4) as : ~하므로(이므로), ~이니까

As it was cold, I nearly caught cold.

I missed Chu-no last night, as I had too much homework yesterday.

2. 명사절을 이끄는 종속 접속사

문장에서 주어, 목적어, 보어 역할을 하는 명사절을 이끈다.

(1) that …인 것, …하다는 것

That he is rude is true.(주어 역할)=It is true that he is rude.

I hope that she is a good woman.(목적어 역할)

The bad news is that there was a car accident.(보어 역할)

(2) whether, if …인지 아닌지

I don't know whether(if) she is a good woman or not.

3. 양보를 나타내는 종속접속사

(1) though, although, even if, even though : (비록) ~일지라도

Though he was poor, he liked to help others.

Basic Test

※ ( ) 안에서 알맞은 말을 고르시오.

01 What are you going to do (if, that) you have 100 dollars?

02 (Because, That) she was ill, she couldn't go hiking.

03 I think (if, that) Mijeong likes you.

※ 다음 밑줄 친 부분에 유의하여 내용이 자연스럽게 이어지도록 연결하시오.

04 Because you didn't study hard, • • ⓐ that you can do it.

05 If Joe invites you to the party, • • ⓑ you couldn't pass the exam.

06 I think • • ⓒ will you go?

※ 우리말과 뜻이 같도록 빈칸에 알맞은 접속사를 쓰시오.

07 네가 몸이 별로 좋지 않다면 일찍 가도 좋다.

__________ you don't feel well, you can leave early.

08 피곤하기 때문에 나는 자고 싶다.

I want to sleep __________ I am tired.

09 나는 네가 그를 사랑한다는 것을 알고 있다.

I know __________ you love him.

※ 우리말과 뜻이 같도록 주어진 단어를 배열하시오.

10 Robert는 나에게 나를 사랑한다고 말했다. (he, me, loved, that)

Robert told me __________.

11 눈이 와서 우리는 산에 올라갈 수 없었다. (it, because, snowed)

We couldn't climb the mountain __________.

12 질문이 있으면 손을 드세요. (questions, if, any, you, have)

Raise your hand __________.

13 나는 네 꿈이 가수가 되는 것이라는 걸 알아. (your, singer, a, to, be, dream, that, is)

I know __________.

Practice Test

01 다음 빈칸에 들어갈 알맞은 말을 고르시오.

_______ you want to get good grades, you should study more.

① If ② Why ③ That
④ But ⑤ Or

02 다음 빈칸에 들어갈 알맞은 말을 고르시오.

The truth is ______ she never loved her husband.

① before ② that ③ because
④ or ⑤ but

03 다음 빈칸에 들어갈 알맞은 말을 고르시오.

A : Why did you stop?
B : _____ I had to answer my phone.

① If ② But ③ Because
④ So ⑤ That

04 다음 빈칸에 들어갈 말이 순서대로 짝지어진 것은?

• Call me _____ you need anything.
• I'm tired _____ I was out late last night.

① because–if ② if–that
③ if–because ④ that–because
⑤ that–if

05 다음 우리말을 영어로 바르게 옮긴 것은?

어머니는 내가 의사가 되길 바라셨다.

① That my mother hoped I was a doctor.
② My mother hoped and I will be a doctor.
③ My mother hopes that would be I a doctor.
④ My mother hoped that I would be a doctor.
⑤ My mother hoped if I would be a doctor.

06 다음 중 어법상 틀린 것은?

① I can't love her because I know Timloves her, too.
② If you want to marry her, tell her you love her.
③ I can't go right now that I'm so busy.
④ Don't tell him that I'm staying here.
⑤ If you decide to help her, you'll get in trouble.

07 다음 중 밑줄 친 부분의 쓰임이 나머지 넷과 다른 것을 고르시오.

① I need your help if you don't mind.
② She wants to know if she can borrow some money from him.
③ If you go away from me, I'll miss you.
④ I can help you if you help me, too.
⑤ Ronald will be excited if you go to the movies with him.

Reading Comprehension

01 다음 글의 내용과 일치하도록 할 때 아래 질문에 대한 답으로 알맞은 것은?

Most bears are big animals. But the sun bear is different from other bears. It is the smallest one. It looks like a big dog. Sun bears stay in trees. During the day, they sleep there. At night, they look for food there. Sun bears eat small animals, birds, bugs, and fruits. They are especially fond of sweet food. They like honey most of all. Every sun bear has a mark. It looks like the sun coming up. Some people think that is how the sun bear got its name.

Why is the smallest bear called 'sun bear'?

① Because of its mark.
② Because of its body size.
③ Because of the food they eat.
④ Because it looks like the sun.
⑤ Because it likes sunny weather.

02 다음 글의 내용과 일치하지 않은 것은?

Early peoples danced to please their gods. Today most people dance just for fun. In fact, it was not until the 12th century that dancing with a partner became popular. In the 19th century, people began dancing the waltz and tango—it was the first time dance partners held each other closely. In the past century dancing styles have changed again. In many cases, set steps have been replaced by solo dancers who make up all their own moves.

① 선조들은 신의 여흥을 돋우기 위해 춤을 췄다.
② 오늘날 춤은 오락을 목적으로 한다.
③ 12세기부터 짝을 이뤄 춤추는 것이 인기를 얻었다.
④ 왈츠는 두 사람이 함께 춘 최초의 춤이다.
⑤ 솔로 무용수들에 의해 춤을 추는 스타일이 변화되었다.

New Words sun bear 반달곰

전치사

› **UNIT 1.** 시간의 전치사
(in, on, at, around, before, after)

› **UNIT 2.** 장소의 전치사
(in, on, at, near, under)

› **UNIT 3.** 그 밖의 전치사
(next to, across from, between and, from to)

UNIT 1 시간의 전치사(in, on, at, around, before, after)

Prestudy

▣ 다음을 해석하시오.

1. She'll visit Japan in September.

____

2. I usually go shopping on Saturday.

____

3. The concert will end around 10 p.m.

____

▣ 다음을 영작하시오.

1. 많은 사람들이 겨울에 스키를 탄다.

____

2. 월요일 몇 시에 회의가 있니? -9시에.

____

3. 방과 후에 뭘 했니?

____

1. **전치사 :** 명사나 대명사 앞에 쓰여서 시간, 장소, 방법, 이유 등을 나타낸다.

(1) in, on, at

	쓰임	예
at	몇시, 몇분, 밤, 정오, 새벽, 시각 등 구체적인 시간	at seven o'clock, at night, at midnight, at noon, at dawn(=at daybreak)
on	날짜, 요일, 특정한 날, 또는 정해진 시간	on Monday, on November 14th, on my birthday, on Christmas Day, on the morning of January 1
in	월, 계절, 년도, 세기, 아침, 오후, 저녁	in July, in Winter, in 2013, in the 20th century, in the morning, in the afternoon, in the evening

Our summer vacation begins in July.

I was born on August 12th.

She left at five o'clock.

The house was fired last night. [그 집에 지난 밤에 불이 났다.] –at lsat night (×)

I am going to meet her next Saturday. [나는 다음 토요일에 그녀를 만날것이다.] –on next Saturday (×)

※ this, that, last, next, all, every, some 등이 시간을 나타내는 단어 앞에 오면 전치사를 생략한다.

(2) around ~ 무렵, ~경에

I like to take a walk around sunrise.

(3) before/after ~전에, ~후에

Wash you hands before lunch.

Let's go to the movies after the meeting.

Basic Test

※ ( ) 안에서 알맞은 말을 고르시오.

01 I was born (in, on) October 1, 1991.

02 My birthday is (on, in) January.

03 Jane watches television (on, in) the evenings.

04 We go to bed late (in, at) night.

※ 다음 우리말의 밑줄 친 부분에 유의하여 빈칸에 알맞은 전치사를 쓰시오.

05 James는 1990년에 고등학생이었다.
James was in high school ________ 1990.

06 방과 후에 같이 공부하는 거 어때?
What about studying together _____ school?

07 나는 오후에는 도서관에서 공부한다.
I study at the library ______ the afternoon.

08 나는 세 시 전에 집에 가야 한다.
I have to go home ________ three o'clock.

※ 다음 우리말에 유의하여 밑줄 친 부분을 바르게 고쳐 쓰시오.

09 그는 10시 무렵에 돌아왔다.
He got back in 10 o'clock.

10 그 가게는 토요일에는 문을 열지 않는다.
The store doesn't open in Saturday.

11 나는 너무 피곤했기 때문에 밤에 잠을 잘 수가 없었다.
I couldn't sleep in night because I was very tired.

※ 다음 문장에서 어법상 틀린 부분을 찾아 바르게 고치시오.

12 I have many classes at the morning.

13 Sumin usually goes to church on nine o'clock.

Practice Test 

※ 다음 빈칸에 들어갈 알맞은 말을 고르시오.

01 Are you going to marry ________ June?

① in ② on ③ at

④ for ⑤ of

02 They promised to meet at the same place ________________ April 11th.

① in ② on ③ to

④ for ⑤ at

03 She'll arrive here __________ 7 o'clock.

① in ② on ③ to

④ for ⑤ at

04 We had a meeting ________ two hours.

① with ② on ③ to

④ for ⑤ at

05 다음 중 어법상 틀린 것을 고르시오.

① She doesn't work on Saturday.

② He had breakfast at 7 a.m.

③ I feel sleepy in spring.

④ Will you see him at school on noon?

⑤ Every amusement park is full of people on Children's Day.

06 다음 중 어법상 틀린 것을 고르시오.

① Roger was born in 1992.

② Kenny bought DVDs for me.

③ You should come back before dinner.

④ What did you do in the morning?

⑤ Laura often walks around at last night.

07 다음 우리말을 영어로 바르게 옮긴 것은?

> 토요일에 영화 보러 가자.

① Let's go to the movies on Saturday.

② Let's go to on Saturday the movies.

③ Let's go to the movies at Saturday.

④ Let's go on Saturday to the movies.

⑤ Let's go to the movies in Saturday.

08 다음 빈칸에 들어갈 말이 순서대로 바르게 짝지어진 것은?

> Yesterday I had an appointment with Suji ___ 7 p.m. But I was late. She was upset, so I had to go to her home _______ night and apologize.

① on－at ② at－at

③ at－in ④ on－on

⑤ at－on

Reading Comprehension

01 브루스 리에 대한 설명중 다음 글의 내용과 일치 하지 않는 것은?

Bruce Lee was a kung fu actor whom many people liked. In fact, he was the movie star who made kung fu famous. In the early 1970's, a Hollywood movie director saw Bruce in a karate championship. He offered to make a movie with him. Bruce Lee made his first successful movie, The Big Boss, in 1971. But the movie that people liked best was Enter the Dragon. It was made in 1973. The movie made him a world star but, unfortunately, it was also his last. He died the same year at the age of thirty-three.

① 쿵푸를 대중에게 널리 알린 영화 스타였다.
② 가라데 선수권 대회를 계기로 영화 배우로 발탁했다.
③ The Big Boss가 그의 첫 흥행 영화 였다.
④ 그의 최고 인기작은 Enter the Dragon이다.
⑤ 영화 Enter the Dragon을 찍는 중에 사망했다.

02 다음 글 (A) 와 (B)의 요지를 바르게 짝지어 진 것은?

(A) I'm worried because I've got a terrible memory. I forget people whom I've met before. I forget where I've put things and I can never remember where I've parked my car. I'm always late in the mornings because I forget to set my alarm clock. Please help me!

(B) You shouldn't worry too much, because there's an easy answer. Why don't you write things down? Keep a list of everything you want to remember. Then you'll begin to remember things. But don't lose the piece of paper!

① (A) 건망증 때문에 고민이에요.
(B) 기억해야 할 일을 적어 두세요.
② (A) 기억력이 점점 나빠져요.
(B) 걱정을 줄이면 나아질 거예요.
③ (A) 기억할 일이 너무 많아요.
(B) 모든 걸 기억하려고 하지 마세요.
④ (A) 기억력을 향상시킬 방법이 없나요?
(B) 한 가지 일에만 집중하세요.
⑤ (A) 오래된 일은 기억이 안나요.
(B) 메모지를 항상 휴대하세요.

UNIT 2 장소의 전치사(in, on, at, near, under)

Prestudy

▣ 다음을 해석하시오.

1. I put the gift in the box.

____

2. I stayed at home all day.

____

3. The hotel is near the airport.

____

▣ 다음을 영작하시오.

1. 그녀는 학교에서 공부하고 있다.

____

2. 벽에 있는 그 그림들을 봐.

____

3. 그 개는 내 책상 밑에 앉아 있다.

____

Key Point

(1) in, at, on

	의미&쓰임	예
at	구체적 개념에서 특정한 지점을 가리키거나 비교적 좁은 장소를 나타낸다. –"~에"	at home, at a party, at school
on	특정 지점의 접촉면과 맞닿아 있는 –"~위에"	on the table, on the wall
in	공간적, 추상적 개념에서 "~안에"라는 의미로 사용되며, 도시나 나라같은 비교적 넓은 장소를 나타낸다.	in the box, in the room, in New York, in Korea

You should be quiet in the library.
Why don't we meet at the cinema?
Let's sit down on the floor.

(2) near : ~근처에

Is there a bus stop near here?

(3) under : ~아래에

Look under the sofa.

(4) 그밖의 전치사

① in front of : ~앞에

Let's meet in front of bookstore.

② with : ~와 함께, ~을 가지고

Can you play tennis with me?

③ by : ~에 의하여, ~을 타고

I go to school by bus.

④ until : ~까지

Ken will stay here until next Monday.

⑤ Over : ~ 바로 저기 위에

Look at thebridge over the river.

⑥ Above : ~보다 높이, ~위에

He lifted his hands above his head.

⑦ Below : ~보다 아래, ~아래

The sun had already sunk below the horizon.

Basic Test

※ (　) 안에서 알맞은 말을 고르시오.

01 Mr. Smith arrived (at, in) Seoul yesterday.

02 The picture hangs (in, on) the wall.

03 Where does she live? –She lives (in, on) Busan.

04 Did you see Max? –He was working (on, at) his office.

※ 우리말과 뜻이 같도록 빈칸에 알맞은 전치사를 쓰시오.

05 그 전등은 식탁 위에 있다.
The lamp is ________ the table.

06 그 고양이는 내 차 근처에서 놀고 있었다.
The cat was playing ____________ my car.

07 Jason은 한국에 살기를 원한다.
Jason wants to live ______________ Korea.

08 나는 네 모자를 책상 위에 두었어.
I put your hat ________________ the desk.

※ 우리말과 뜻이 같도록 다음 밑줄 친 부분을 바르게 고쳐 쓰시오.

9 너는 밖에 나가고 싶니, 아니면 집에 있고 싶니?
Do you want to go out or stay <u>on</u> home?

10 파리에는 볼 것이 많이 있다.
There are many things to see <u>on</u> Paris.

11 벽에 있는 저 그림은 누가 그렸니?
Who painted the picture <u>in</u> the wall?

※ 다음 밑줄 친 부분에 유의하여 우리말로 옮기시오.

12 There are many health clubs <u>near</u> the stadium.

____

13 I met Emma <u>at</u> Jim's birthday party.

____

New Words hang 걸다, 걸려있다

Practice Test

※ 다음 빈칸에 들어갈 알맞은 말을 고르시오.

01 Is there a bank ______________ here?

① by　② at
③ under　④ near
⑤ on

02 I was waiting for you _____ the bus stop.

① in　② at
③ over　④ on
⑤ upon

03 There were so many new people ________ the party last night.

① in　② at
③ under　④ near
⑤ into

04 다음 중 어법상 <u>틀린</u> 것을 고르시오.

① It's cool under the trees.
② She'll stay in Korea.
③ There's no bank near here.
④ Dr. Kim was in the office.
⑤ I didn't stay on home.

05 다음 중 어법상 <u>틀린</u> 것을 고르시오.

① They live in this apartment.
② She gave many movie DVDs at me.
③ He found the keys under the sofa.
④ Yujin arrived at the airport at three.
⑤ In his house, there are no pictures on the walls.

06 다음 우리말을 영어로 바르게 옮긴 것을 고르시오.

> 그는 아쿠아리움 앞에 있어.

① He is at the Aquarium.
② He is in the Aquarium.
③ He is in front of the Aquarium.
④ He is on the Aquarium.
⑤ He is in front the Aquarium of.

07 다음 우리말을 영어로 바르게 옮긴 것을 고르시오.

> 그녀는 다음 주 월요일까지 파리에 머물 것이다.

① She'll stay in Paris in next Monday.
② She'll stay in Paris on next Monday.
③ She'll stay in Paris at next Monday.
④ She'll stay in Paris until next Monday.
⑤ She'll stay in Paris from next Monday.

Reading Comprehension

01 다음 글의 주제로 가장 적절한 것은?

We are living in the "Information Ages." Computers are becoming more and more necessary in our society. I think computer programming will be a good job in the future. I'll study hard and learn all about computers. Someday I will be a good computer programmer.

① 현대 사회의 특징
② 컴퓨터의 필요성
③ 나의 미래 희망
④ 정보 사회와 컴퓨터
⑤ 컴퓨터 학과의 장래성

※ 다음 글을 읽고 물음에 답하시오.

Baby elephants get extra care. A baby elephant's mother looks after it. The baby is also looked after by an aunt. Lions and tigers won't go after a fully grown elephant. But they will go after a baby. When the elephants walk along a trail, the mother elephant goes first. The aunt comes last. And <u>the baby is in the middle</u>. The baby elephants walk between the mother and the aunt so that lions and tigers can't get to it.

02 밑줄 친 <u>the baby is in the middle</u>. 의 이유로 가장 적절한 것은?

① 맹수들의 접근을 막기 위하여
② 무리에서 뒤쳐지지 않기 위하여
③ 힘들 때 엄마 젖을 먹기 위하여
④ 도중에 길을 잃어버리지 않기 위하여
⑤ 어미들 틈에서 노는 것을 좋아하므로

03 이 글의 주된 내용을 가장 잘 나타낸 것은?

① Baby elephants get extra care.
② A baby elephant's mother looks after it.
③ The baby is also looked after by an aunt.
④ Lions and tigers will go after a baby elephant.
⑤ The baby walks between the mother and the aunt.

UNIT 3 그 밖의 전치사(next to, across from, between and, from to)

Prestudy

▣ 다음을 해석하시오.

1. Who is standing next to you?

2. The baseball stadium is across from the park.

3. The child is between his mother and his father.

4. Sho took a trip from LA to New York.

▣ 다음을 영작하시오.

1. 버거킹은 백화점 옆에 있다.

2. 그 건물은 서울역 맞은편에 있다.

3. 의자는 TV와 컴퓨터 사이에 있다.

Key Point

(1) next to : ~옆에, 곁에

The pretty girl sat next to me. So I was happy.

(2) across from : 맞은편에(=opposite)

There's an Internet cafe across from my house.

(3) between A and B : A와 B사이에

The shop is between the bank and the bakery.

(4) from A to B : A부터 B까지

Jane works from 9 a.m. to 5 p.m.

Basic Test

※ ( ) 안에서 알맞은 말을 고르시오.

01 I want to take a trip (between, from) January to March.

02 The eraser is next (to, with) my book.

03 They were sitting between me (and, from) him.

04 His house is across (to, from) the grocery store.

05 I had a seat between the lady (and, to) the young boy.

※ 우리말과 뜻이 같도록 밑줄 친 부분을 바르게 고쳐 쓰시오.

06 그 고양이는 창문 옆에서 자고 있었다.
The cat was sleeping beside to the window.

07 그 슈퍼마켓은 학교와 우리 집 사이에 있다.
The supermarket is in the middle the school and my house.

08 그는 그 빌딩 건너편에 서 있었다.
He was standing over from the building.

09 그는 비행기로 런던에서 서울까지 갔다.
He flied for London to Seoul.

※ 우리말과 뜻이 같도록 빈칸에 알맞은 말을 쓰시오.

10 Tom은 내 옆에 앉아 있었다.
Tom was sitting ________________ to me.

11 Brown 박사의 방은 그 사무실 맞은편이다.
Dr. Brown's room is ____ from the office.

12 내 차는 도서관과 상점 사이에 있다.
My car is __________ the library and the store.

13 나는 매일 아침 공원에서 학교까지 산책한다.
I take a walk _______ the park __________ the school every morning.

New Words grocery store 식료품점

Practice Test

※ 다음 빈칸에 들어갈 알맞은 말을 고르시오.

01 Jane's desk is _______ the sofa and the bookshelf.

① at ② around ③ between
④ among ⑤ about

02 The marathon course is _______ Seoul to Suwon.

① at ② around ③ between
④ from ⑤ across

03 There's a fire station _____ to the school. You can find it easily.

① around ② next ③ about
④ between ⑤ across

04 다음 중 어법상 틀린 것을 고르시오.

① She'll come back between 7 p.m. and 8 p.m.
② He's sitting right across from Sally.
③ The river flows next to the building.
④ You can choose among these two colors.
⑤ We're open from 9 a.m. to 10 p.m.

05 다음 중 어법상 틀린 것을 고르시오.

① I waited for him at the bank.
② Don't look at me!
③ I like to listen in hip hop.
④ It's across from City Hall.
⑤ Go and sit next to Tom.

06 다음 우리말을 영어로 바르게 옮긴 것을 고르시오.

그것은 우리 둘 사이의 비밀이야.

① It's a secret between the of two us.
② It's a between secret the two of us.
③ It's a secret between the two of us.
④ It's a secret among the two of us.
⑤ It's a secret between of the two us.

07 다음 우리말을 영어로 바르게 옮긴 것을 고르시오.

그 박물관은 오전 10시부터 오후 5시까지만 연다.

① The museum is only open next 10 a.m. to 5 p.m.
② The museum is only open from 10 a.m. to 5 p.m.
③ The museum is only open from 10 a.m. and 5 p.m.
④ The museum is only open between 10 a.m. to 5 p.m.
⑤ The museum is from 10 a.m. to 5 p.m. open.

Reading Comprehension

01 다음 글의 제목으로 알맞은 것은?

We and our animal brothers are part of nature. The whole animal kingdom, in which the human beings are dear members, depends a lot on nature. Nature gives us food, clothes, and shelter. It's just like a mother taking care of her children. That is why people speak of Mother Nature. Certainly we should be nice to our mother.

① Love of Nature
② The Animal Kingdom
③ The Human Beings
④ Mother and Children
⑤ Food, Clothes, and Shelter

※ 다음 글을 읽고 물음에 답하시오.

Tom and John went for a walk. On the road they saw a wallet. There was a lot of money in it. Tom picked it up. "I have a wallet full of money," he said and put it in his pocket. "Don't say 'I', say 'we'" said John. "It is our wallet, not yours." "No," said Tom. "I saw it first. It is my wallet." Then a man ran up to them. "That is my wallet," he said. "I saw you. You are a thief. I'll get a policeman." "What shall we do?" said Tom to John. "Don't say 'we', say 'I'" said John. "You picked up the wallet, not I."

02 밑줄 친 "Don't say 'we', say 'I'" 를 통해 John이 말하고자 하는 의도로 가장 적절한 것은?

① 나는 나의 지갑을 훔치지 않았다.
② 지갑 속의 돈을 주인에게 돌려주자.
③ 주운 지갑에 대해서 나는 책임이 없다.
④ 나는 이제 더 이상 너의 친구가 아니다.
⑤ 주운 지갑 속의 돈을 나누어 갖자.

03 이 일을 겪고 난 뒤 Tom이 느끼는 심정으로 가장 적절한 것은?

① 기쁘다 ② 억울하다
③ 겸연쩍다 ④ 슬프다
⑤ 만족스럽다

군 부대의 단위 해설

Listen! Our squad shares a single fate, we shall move together at all costs.
(잘 들어! 우리 분대는 운명을 같이한다. 우리는 어떠한 경우에도 같이 움직인다.)

➡ squad는 병사 10명과 하사 2명으로 구성된 '분대'를 뜻합니다. 미군에서 사용되는 용어입니다. 영국군에서는 section이라는 표현이 '분대'를 의미하며 이 단어 section이 미군에서 사용될 경우 '반(半) 분대'를 뜻합니다.

Helicopters thrummed in to deposit entire platoons of infantrymen, and armored personnel carriers rumbled through the mire at 450 yards an hour in hopes of pinning the enemy troop to the sticking point.
(헬리콥터들은 윙윙거리며 소대단위의 보병들을 계속 수송해 왔고 병력 수송용 장갑차들은 적 부대를 옴짝달싹 하지 못하게 궁지에 몰아 넣을 심산으로 시속 450 야드의 속도로 덜거덕거리며 진창을 헤쳐 나갔다.)

➡ platoon은 '소대'입니다. 2개 분대 이상으로 편성된 군 단위입니다. 규모는 30명 정도입니다.

There must be more reinforcements to keep the isolated company from annihilation.
(고립된 중대가 전멸하는 것을 방지하기 위해 더 많은 보충 병력이 지원돼야 합니다.)

➡ company는 4개의 소대로 편성되어 대위(大尉)가 지휘하는 '중대'입니다. 보통 120명의 군사가 있습니다. 참고로 보병 중대는 an infantry company, 기병 중대는 a troop, 포병 중대는 a battery, 비행 중대는 영국의 경우 a squadron이라 합니다.

The fifth battalion will go to Iraq to supplement the losses.
(제 5대대는 병력 손실을 보충하기 위해 이라크로 갈 것이다.)

➡ battalion은 '대대'를 뜻합니다. 보통 500에서 1000명으로 구성된 꽤 큰 규모의 부대입니다.

The regiment suffered a deterioration of discipline as the newly employed colonel displayed serious inconsistency in his leadership.
(새로 임용된 대령의 심각하게 일관성 없는 지도력에 의해 연대의 기강이 해이해졌다.)

➡ regiment는 '연대'입니다. 보통 연대는 3개의 보병 대대로 이루어져 있습니다.

Under the United Nations flag, the British brigade, including Australian, Canadian and New Zealand troops, participated in the Kapyong battle against North Koreans backed by Chinese forces in 1951.
(유엔의 깃발 아래 호주, 캐나다 그리고 뉴질랜드의 부대가 포함된 영국의 여단이 북한군과 그들을 지원하는 중국군

에 맞서 1951년에 있었던 가평 전투에 참여했었다.)

➡ brigade는 보통 2개 이상의 연대로 이루어진 '여단'입니다. 여단은 보통 장성에 의해 지휘됩니다.

➡ 마지막으로 이 여단이 2개 이상 모여 20,000명 규모의 division(사단)이 되고, 다시 이 division(사단)이 2개 이상 모여 corps(군단)을 이룹니다.

정답 및 해설

CHAPTER 1 be동사와 인칭대명사

| UNIT 1 | be동사와 인칭대명사

■ Prestudy

▣ 다음을 영작하시오.

1. I am a singer.
2. He's handsome.
3. We were in Busan yester

• Basic Test

1. is 2. are 3. were 4. was 5. am 6. is 7. are 8. was 9. were 10. is, He is 11. They are 12. We are 13. was, It was

• Practice Test

1. ③ 2. ⑤ 3. ④ 4. ② 5. ② 6. ② 7. ③

• Reading Comprehension

1. ③ 2. ④

| UNIT 2 | be동사의 부정문과 의문문

■ Prestudy

▣ 다음을 영작하시오.

1. He isn't hungry.
2. Are they baseball players?

• Basic Test

1. Are 2. are not 3. Was 4. is not 5. Are 6. aren't 7. was 8. weren't 9. Minhee was not[wasn't] at the dance party./Was Minhee at the dance party? 10. Mike and Susie are not[aren't] from England./Are Mike and Susie from England? 11. They are not from China 12. Was he in his office last Friday 13. Is she your sister

• Practice Test

1. ① 2. ① 3. ③ 4. ③ 5. ④ 6. ④

• Reading Comprehension

1. ③ 2. ③

| UNIT 3 | 인칭대명사의 격변화

■ Prestudy

▣ 다음을 영작하시오.

1. He is clever.
2. her smile is beautiful
3. My parents love me.
4. The letter on the desk is yours.

• Basic Test

1. You 2. theirs 3. hers 4. us, him 5. mine 6. your 7. him 8. She 9. my, her 10. your, yours 11. our, We, her 12. Mine is old 13. She loves him, too

• Practice Test

1. ③ 2. ② 3. ④ 4. ① 5. ④ 6. ② 7. ②

• Reading Comprehension

1. ③ 2. ①

CHAPTER 2 일반동사

| UNIT 1 | 일반동사의 의미와 형태

■ Prestudy

▣ 다음을 영작하시오.

1. I need your help
2. You speak English well
3. She likes music.

• Basic Test

1. has 2. makes 3. speaks 4. does 5. goes 6. kisses 7. fixes 8. cries 9. helps 10. takes 11. plays 12. drinks orange juice every morning 13. has a lot of work

• Practice Test

1. ④ 2. ③ 3. ⑤ 4. ③ 5. ③
6. ④ 7. ①

• Reading Comprehension

1. ③ 2. ②

| UNIT 2 | 일반동사의 과거형

■ Prestudy

▣ 다음을 영작하시오.

1. I watched movie last night.
2. I came back home after a trip to Europe.

• Basic Test

1. met 2. went 3. left 4. read 5. had 6. watched 7. made 8. came 9. enjoyed 10. arrived 11. 스미스 씨는 오늘 아침에 일찍 일어났다. 12. 아빠는 5년 전에 담배를 끊으셨다. 13. 한국 축구팀은 2002 월드컵에서 경기를 매우 잘했다.

• Practice Test

1. ③ 2. ② 3. ⑤ 4. ① 5. ① 6. ④
7. ①

• Reading Comprehension

1. ④ 2. ③

| UNIT 3 | 일반동사의 부정문

■ Prestudy

▣ 다음을 영작하시오.

1. I don't tell lies.
2. She doesn't like him
3. They didn't know my name.

• Basic Test

1. does not 2. doesn't go 3. don't read 4. like 5. didn't 6. doesn't[didn't] 7. doesn't drive 8. don't work 9. don't like 10. didn't do 11. doesn't rise 12. didn't brush 13 don't[-didn't] watch

• Practice Test

1. ① 2. ③ 3. ② 4. ② 5. ⑤ 6. ③
7. ④

• Reading Comprehension

1. ③ 2. ②

| UNIT 4 | 일반동사의 의문문

■ Prestudy

▣ 다음을 영작하시오.

1. Do you like sports?–Yes, I do.
2. Does he wear glasses?–No, he doesn't.
3. Did she catch a cold?–Yes, she did.

• Basic Test

1. Does 2. Do 3. Does 4. Do 5. Do you visit 6. Did Jennifer see 7. Do they live 8. Does the store sell 9. Do you 10. Does Jane 11. Do Harry and Sally 12. he does 13. she doesn't

• Practice Test

1. ④ 2. ③ 3. ⑤ 4. ④ 5. ⑤ 6. ①

• Reading Comprehension

1. ④ 2. ①

CHAPTER 3 일반동사

| UNIT 1 | 명사의 단수와 복수

■ Prestudy

▣ 다음을 영작하시오.

1. I have two dogs.
2. There were three women on the talk show.
3. He drinks a cup of coffee in the morning.

• Basic Test

1. boxes 2. gentlemen 3. ladies 4. two cups of coffee 5. cup 6. glasses 7. bottle 8. piece 9. two sheet of papers→two sheets of paper 10. potatos→potatoes 11. childs→children 12. children, girls[boys], boys[girls] 13. leaves, trees

• Practice Test

1. ③ 2. ② 3. ④ 4. ① 5. ④ 6. ③
7. ⑤

• Reading Comprehension

1. ④ 2. ②

| UNIT 2 | 관사

■ Prestudy

▣ 다음을 영작하시오.

1. She is a designer.
2. I bought a pet dog. The dog was very cute.
3. Let's meet here after lunch.

• Basic Test

1. any money 2. an 3. the 4. baseball 5. a→an 6. sheeps→sheep 7. a salt→the salt 8. A sun→The sun 9. × 10. a 11. a, a, The, the 12. 사과를 하루에 하나씩 먹으면 의사를 멀리하게 된다[건강해진다]. 13. 수잔은 주말에 기타를 친다.

• Practice Test

1. ① 2. ③ 3. ② 4. ② 5. ⑤ 6. ② 7. ④ 8. ③

• Reading Comprehension

1. ④ 2. ①

CHAPTER 4 의문사

| UNIT 1 | who, what, which

■ Prestudy

▣ 다음을 영작하시오.

1. Who is that cool guy?
2. What is your favorite TV program?
3. Which do you like better, action movies or comedy movies?

• Basic Test

1. What 2. Who 3. Which 4. Who 5. What 6. What 7. Which 8. What 9. Who 10. ① 11. ③ 12. ⑥ 13. ②

• Practice Test

1. ④ 2. ② 3. ① 4. ③ 5. ④ 6. ④ 7. ③

• Reading Comprehension

1. ⑤ 2. ②

| Unit 2 | when, where, why

■ Prestudy

▣ 다음을 영작하시오.

1. When is his birthday?
2. Where is post office around here?
3. Why do you like your girl friend?

• Basic Test

1. Where 2. When 3. When 4. Why 5. Where 6. ③ 7. ① 8. ④ 9. Where 10. When 11. Why didn't he come to the party

• Practice Test

1. ③ 2. ④ 3. ④ 4. ③ 5. ⑤ 6. ③

• Reading Comprehension

1. ④ 2. ③

| UNIT 3 | how

■ Prestudy

▣ 다음을 영작하시오.

1. How is the weather in Canada?
2. How did you know her name?
3. How many pens do you need?

• Basic Test

1. How 2. What 3. How 4. much 5. How 6. How 7. many 8. ③ 9. ② 10. ⑤ 11. ④

• Practice Test

1. ① 2. ③ 3. ② 4. ④ 5. ③ 6. ④

• Reading Comprehension

1. ③ 2. ②

CHAPTER 5 동사의 시제

| UNIT 1 | 현재시제와 과거시제

■ Prestudy

▣ 다음을 영작하시오.

1. He doesn't eat fast food.
2. I had a good time with him yesterday.

• Basic Test

1. went 2. began 3. traveled 4. stayed 5. swam 6. came 7. read 8. bought 9. died 10. rises 11. invented 12. learned 13. won

• Practice Test

1. ② 2. ④ 3. ② 4. ⑤ 5. ⑤ 6. ③ 7. ①

• Reading Comprehension

1. ④ 2. ⑤

| UNIT 2 | 진행시제

■ Prestudy

▣ 다음을 영작하시오.

1. My boyfriend is dancing on the stage.
2. Jim was making his blog.
3. Are you listening to me?

• Basic Test

1. was 2. was taking 3. is cooking 4. was 5. was 6. were 7. were 8. staying 9. ② 10. ③ 11. ④ 12. 그들은 교실에서 노래를 하고 있었다. 13. 그녀는 '반지의 제왕'을 세 번이나 읽었다.

• Practice Test

1. ④ 2. ⑤ 3. ② 4. ③ 5. ③ 6. ① 7. ③

• Reading Comprehension

1. ③ 2. ② 3. ①

CHAPTER 6 조동사

| UNIT 1 | will, be going to

■ Prestudy

▣ 다음을 영작하시오.

1. I will not go there with you.
2. The meeting will begin at 3 p.m.
3. We are going to miss you.

• Basic Test

1. will 2. will 3. is going to 4. buy 5. will not go 6. will 7. won't 8. will not go to 9. are you going to do this weekend 10. will be fine tomorrow 11. 그는 훌륭한 예술가[화가]가 될 거야. 12. 우리는 토요일에 영화 보러 갈 거야. 13. 여기에 얼마나 오래 머무를 거야?

• Practice Test

1. ⑤ 2. ② 3. ① 4. ④ 5. ③ 6. ③ 7. ②

• Reading Comprehension

1. ② 2. ④ 3. ④

| UNIT 2 | can, may

■ Prestudy

▣ 다음을 영작하시오.

1. I can cook pizza.
2. May I help you?
3. It may rain this evening.

• Basic Test

1. to sing 2. may 3. can't go 4. Can 5. cannot walk very well 6. may go out and play basketball 7. is able to play chess 8. is able 9. cook 10. help 11. 마크는 한국어를 조금 말한다. 그는 아마도 한국어를 배우고 있는 것 같다. 12. 사라 장은 바이올린을 아주 잘 연주할 수 있다. 13. 저를 좀 도와주시겠습니까?

• Practice Test

1. ③ 2. ① 3. ④ 4. ② 5. ④ 6. ④ 7. ⑤

• Reading Comprehension

1. ② 2. ⑤

| UNIT 3 | must, have to, should

■ Prestudy

▣ 다음을 영작하시오.

1. You must solve this problem.
2. He has to lose weight for his health.
3. You should not walk on the grass.

• Basic Test

1. must 2. must 3. don't have to 4. must 5. clean 6. must 7. should not watch 8. see 9. must exchange it 10. should not eat too much 11. 비가 많이 온다. 너는 집에 있어야 해. 12. 빌게이츠는 마이크로소프트를 소유하고 있다. 그는 부자임에 틀림없다. 13. 너는 일요일에는 오지 않아도 돼.

• Practice Test

1. ③ 2. ④ 3. ⑤ 4. ② 5. ③ 6. ①

• Reading Comprehension

1. ④ 2. ④

CHAPTER 7 문장의 형태

| UNIT 1 | 주어+동사+(수식어)

■ Prestudy

▣ 다음을 영작하시오.

1. There is a vase on the table.
2. Here comes the bus.
3. He goes to school at eight.

• Basic Test

1. ⑤ 2. ② 3. ③ 4. ⑤ 5. ④ 6. ⑤ 7. ④

• Practice Test

1. ⑤ 2. ⑤ 3. ① 4. ③ 5. ⑤ 6. ④

• Reading Comprehension

1. ⑤ 2. ② 3. ⑤

| UNIT 2 | 감각동사+형용사

■ Prestudy

▣ 다음을 영작하시오.

1. She looks young.
2. This fresh fruit tastes good.
3. He came running to meet us.

• Basic Test

1. fresh 2. tall 3. cold 4. soft 5. lovely 6. good 7. exciting 8. feels 9. looked 10. tastes 11. 지수는 아파 보인다. 12. 수프가 맛이 짜다. 13. 너의 거실에 있는 카펫은 감촉이 부드럽다.

• Practice Test

1. ③ 2. ⑤ 3. ① 4. ⑤ 5. ② 6. ① 7. ④

• Reading Comprehension

1. ③ 2. ② 3. ③

| UNIT 3 | 동사+목적어

■ Prestudy

▣ 다음을 영작하시오.

1. I hope to see you again.
2. I think (that) she is beautiful.
3. He ate delicious pizza.

• Basic Test

1. ① 2. ③ 3. ② 4. ③ 5. ⑤ 6. ②

• Practice Test

1. ④ 2. ④ 3. ③ 4. ③ 5. ⑤ 6. ②

• Reading Comprehension

1. ② 2. ③ 3. ①

| UNIT 4 | 수여동사+간접목적어+직접목적어

■ Prestudy

▣ 다음을 영작하시오.

1. James teaches us English.
2. I showed my picture to her.
3. Can you get me apple juice?

• Basic Test

1. gave 2. bought 3. give 4. us math 5. my friend an email 6. her children some cookies 7. ③ 8. ③ 9. ② 10. gave me a DVD player 11. made lunch for her 12. buy his son a pair of skates 13. show me the way to the bus stop

• Practice Test

1. ③ 2. ② 3. ① 4. ④ 5. ④ 6. ⑤ 7. ④

• Reading Comprehension

1. ④ 2. ② 3. ①

| UNIT 5 | 동사+목적어+목적보어

■ Prestudy

▣ 다음을 영작하시오.

1. His father made him a doctor.
2. Your smile makes me happy.
3. She wants me to study abroad.

• Basic Test

1. to drive 2. laugh 3. to play 4. clean 5. to go 6. do 7. to speak 8. rich 9. wanted me to work hard 10. makes her happy 11. feel our car move 12. 나는 그들이 나와 함께 야구했으면 해. 13. 그 소식은 나를 슬프게 만들었다.

• Practice Test

1. ③ 2. ③ 3. ① 4. ⑤ 5. ② 6. ④ 7. ③ 8. ④

• Reading Comprehension

1. ④ 2. ④ 3. ③

CHAPTER 8 대명사

| UNIT1 | 부정대명사 one, some, any

■ Prestudy

▣ 다음을 영작하시오.

1. Do you have a pet?–Yes, I have one.
2. There are some books on the table.

• Basic Test

1. one 2. it 3. one 4. one 5. others 6. the other 7. one 8. others 9. it 10. it 11. one 12. ones 13. panda 14. notebook computer 15. some→a 또는 one→ones 16. One→It 17. some→any

• Practice Test

1. ④ 2. ⑤ 3. ③ 4. ③ 5. ④ 6. ⑤ 7. ③

• Reading Comprehension

1. ④ 2. ⑤

| UNIT 2 | 재귀대명사

■ Prestudy

▣ 다음을 영작하시오.

1. May I introduce myself?
2. He is talking to himself.
3. I'll do it myself.

• Basic Test

1. herself 2. yourself 3. ourselves 4. myself 5. ourselves 6. himself 7. myself 8. yourself 9. himself 10. itself 11. 그가 직접 그의 방을 청소해야 한다. 12. 팀은 거울로 자신의 모습을 보았다. 13. 때때로 나는 혼잣말을 한다.

• Practice Test

1. ① 2. ③ 3. ② 4. ① 5. ③ 6. ④ 7. ⑤ 8. ①

• Reading Comprehension

1. ④ 2. ② 3. ③ 4. ①

CHAPTER 9 형용사와 부사

| UNIT 1 | 형용사와 부사의 역할

■ Prestudy

▣ 다음을 영작하시오.

1. We live in a big house.
2. My sister is very pretty.
3. I often go to the library.

• Basic Test

1. happy 2. kind 3. something cold 4. diligently 5. smart 6. clearly 7. often plays 8. fast[quickly] 9. slowly 10. interesting 11. Listen to me carefully 12. always smiles at me 13. is lovely today

• Practice Test

1. ③ 2. ④ 3. ⑤ 4. ② 5. ③ 6. ③ 7. ④

• Reading Comprehension

1. ① 2. ④ 3. ② 4. ②

| UNIT 2 | 비교 표현(원급, 비교급, 최상급)

■ Prestudy

▣ 다음을 영작하시오.

1. He's as tall as Brian.
2. She looks older than twenty.
3. This is the oldest building on this street.

• Basic Test

1. heavier 2. happier 3. easier 4. more exciting 5. taller 6. worse 7. smart 8. prettier 9. faster 10. larger 11. older than I 12. the best swimmer 13. the most exciting

• Practice Test

1. ② 2. ④ 3. ③ 4. ③ 5. ④ 6. ④ 7. ①

• Reading Comprehension

1. ② 2. ① 3. ②

CHAPTER 10 접속사

| UNIT 1 | 등위접속사 and, but, or

■ Prestudy

▣ 다음을 영작하시오.

1. Polar bears are tall white.
2. Jane likes science, but I like math.
3. Do you want iced tea or hot tea?

• Basic Test

1. and 2. but 3. or 4. but 5. and 6. but 7. or 8. and 9. or 10. but 11. 민희와 나는 좋은 친구 사이이다. 12. 가서 그를 돕자! 13. 차나 커피 좀 드시겠어요?

• Practice Test

1. ② 2. ③ 3. ① 4. ③ 5. ③ 6. ④ 7. ①

• Reading Comprehension

1. ④ 2. ④ 3. ②

| UNIT 2 | 종속접속사 when, before, after

■ Prestudy

▣ 다음을 영작하시오.

1. I had a dog when I was a child.
2. Be careful before you cross the street.
3. After they won the game, they had a party.

• Basic Test

1. ⓒ 2. ⓐ 3. ⓑ 4. When 5. when 6. Before 7. after 8. before 9. When 10. before 11. after 12. 나는 커서 변호사가 될 것이다. 13. 그가 나를 떠난 후 나는 그를 그리워했다.

• Practice Test

1. ④ 2. ④ 3. ① 4. ③ 5. ④ 6. ②

• Reading Comprehension

1. ④ 2. ③

| UNIT 3 | 종속접속사 because, if, that

■ Prestudy

▣ 다음을 영작하시오.

1. She closed the window because it was cold.
2. If I win the lottery, I'll travel around the world.
3. I think (that) you can do it.

• Basic Test

1. if 2. Because 3. that 4. ⓑ 5. ⓒ 6. ⓐ 7. If 8. because 9. that 10. that he loved me 11. because it snowed 12. if you have any questions 13. that your dream is to be a singer

• Practice Test

1. ① 2. ② 3. ③ 4. ③ 5. ④ 6. ③ 7. ②

• Reading Comprehension

1. ① 2. ④

CHAPTER 11 전치사

| UNIT 1| 시간의 전치사(in, on, at, around, efore, after)

■ Prestudy

▣ 다음을 영작하시오.

1. Many people ski in winter.
2. What time do you have a meeting on Monday?—At nine.
3. What did you do after school?

• Basic Test

1. on 2. in 3. in 4. at 5. in 6. after 7. in 8. before 9. around[about] 10. on 11. at 12. at→in 13. on→at

• Practice Test

1. ① 2. ② 3. ⑤ 4. ④ 5. ④ 6. ⑤ 7. ① 8. ②

• Reading Comprehension

1. ⑤ 2. ①

| UNIT 2 | 장소의 전치사(in, on, at, near, under)

■ Prestudy

▣ 다음을 영작하시오.

1. She is studying at school.
2. Look at the pictures on the wall.
3. The dog sits under my desk.

• Basic Test

1. in 2. on 3. in 4. at 5. on 6. near 7. in 8. on 9. at 10. in 11. on 12. 경기장 근처에는 헬스클럽이 많이 있다. 13. 나는 짐의 생일파티에서 엠마를 만났다.

• Practice Test

1. ④ 2. ② 3. ② 4. ⑤ 5. ② 6. ③ 7. ④

• Reading Comprehension

1. ③ 2. ① 3. ①

| UNIT 3 | 그 밖의 전치사(next to, across from, between and, from to)

■ Prestudy

▣ 다음을 영작하시오.

1. Burger King is next to the department store.
2. The building is across from Seoul Station.
3. The chair is between the TV and the computer.

• Basic Test

1. from 2. to 3. and 4. from 5. and 6. next 7. between 8. across 9. from 10. next 11. across 12. between 13. from, to

• Practice Test

1. ③ 2. ④ 3. ② 4. ④ 5. ③ 6. ③ 7. ②

• Reading Comprehension

1. ① 2. ③ 3. ②

군사영어 약어 모음

A

AA : Air to Air (공대공)
AA : Assembly Area (집결지)
AAA : Anti-Aircraft Artillery (대공포병)
AAGS : Army Air Ground System (육군공지체계)
AAM : Air to Air Missile (공대공 미사일)
ABC : Atomic, Biological, Chemical (화생방)
ABCCC : Airborne Battlefield Command and Control Center (전장 공중지휘 통제본부)
ACC : Air Component Command (공군 구성군 사령부)
ACC : Airspace Control Center (공역통제본부)
ACE : Airspace Control Element (공역 통제반)
ACE : Airspace Coordination Element (공역 협조반)
ACO : Airspace Control Order (공역통제명령)
ACR : Armored Cavalry Regiment (기갑수색연대)
AD : Active Duty (현역)
ADA : Air Defence Artillery (방공포병)
ADAM : Area Denial Artillery Munition (대인용지뢰탄)
ADC : Aide-De-Camp (전속부관)
ADD : Agency for Defense Development (국방과학연구소)
ADIZ : Air Defence Identification Zone (방공식별구역)
ADOA : Air Defence Operation Zone (방공작전구역)
AFAC : Airborne Forward Air Controller (공중전방항공통제관)
AFKN : American Forces Korea Network (주한 미군 방송)
AFV : Armored Fighting Vehicle (장갑전투차량)
AGM : Air to Ground Missile (공대지 미사일)
AGOS : Air Ground Operations System/School (공지작전체계/학교)
AH : Attack Helicopter (공격용 헬기)
AI : Air Interdiction (항공차단)
AIDE (부관, 보좌관)
AIM : Air Intercept Missile (공중 요격 미사일)
AKAC : Allied Crypto-graphic Operational Code or Cipher (연합작전용 음어)
ALB : Air Land Battle (공지전투)
ALCC : Air Lift Control Center (공수통제본부)
ALO : Air (Army) Liaison Officer (공군(육군) 연락 장교)
AM : Amplitude Modulation (진폭 변조)
AMC : Airborne Mission Commander (공중작전 임무 지휘관)
AMLS : Airspace Liaison Company (공역통제 연락반)
AMRAAM : Advanced Medium Range Air to Air Missile (향상된 중거리 공대공 미사일)
ANGLICO : Air Naval Gunfire Liaison Company (항공 함포 연락중대)
AO : Area of Operation (작전지역)

AOA : Amphibious Objective Area (상륙목표지역)

AOB : Air Order of Battle (공군전투서열)

AOC : Airbase Operation Center (항공 작전본부)

AOP : Aerial Observation Post (공중감시소/대공관측소)

AP : Ammunition Point (탄약분배소)

AP : Anti Personnel (대인)

AP : Armor Piercing (철갑탄)

AP : Auto Pilot (자동 조종 장치)

APC : Armed Personnel Vehicle (인원수송차량)

APFSDS : Armor-Piercing Fin-Stabilized Discarding Sabot (날개안정식 철갑 형성작약)

APPCON : Approach Control (접근관제)

AR : Automatic Rifle (자동화기)

ARS : Automatic Replying System (자동응답체계)

ARTCC : Air Route Traffic Control Center (항로교통 관제본부)

ASEAN : Association of Southeast Asian Nation (동남아 국가연합)

ASG : Area Support Group (지역지원단)

ASOC : Air Support Operation Center (항공지원 작전본부)

ASP : Ammunition Supply Point (탄약보급소)

ASR : Ammunition Supply Rate (탄약보급률)

ASRAAM : Advanced Short Range Air to Air Missile (신형단거리 공대공 미사일)

ATACMS : Army Tactical Missile System (육군 전술유도탄 체계)

ATBM : Anti Tactical Ballistic Missile (대 전술탄도탄)

ATCC : Air Traffic Control Center (항공교통 관제본부)

ATF : Amphibious Task Force (상륙기동부대)

ATP : Ammunition Transfer Point (탄약 전환보급소)

ATP : Ammunition Transfer Point (탄약전환보급소)

ATT : Army Training Test (육군 훈련 시험)

AUM : Air to Underwater Missile (공대수중유도탄)

AV : Aviation (항공)

AW : Automatic Weapons (자동화기)

AWACS : Airborne Warning and Control System (공중경보 및 통제체제)

B

BAI : Battlefield Air Interdiction (전장항공차단)

BCD : Battle Coordination Detachment (전투협조처)

BCE : Battle Coordination Element (전투협조반)

BCTP : Battle Command Training Program (전투지휘 훈련계획)

BDA : Battle Damage Assessment (전투피해평가)

BDA : Bomb Damage Assesment (폭격피해판정)

BDU : Bomb Dummy Unit (모의탄)

BECS : Battlefield Electric CEOI System (전장통신전자운영체계)

BEQ : Bachelor Enlisted Quarters (독신하사관숙소)

BHL : Battle Handover Line (전투이양선)

BL : Basic Load (Ammunition) (기본휴대량)

BM : Ballistic Missile (탄도 미사일)

BMD, BMP : Name of Russian Light Armored Vehicle (러시아제 경장갑차량)

BMNT : Beginning of Morning Nautical Twilight (행상박명초)

BOD : Biochemical Oxygen Demand (생화학 적 산소요구량)

BOQ : Bachelor Officers' Quarters (독신장교숙소)

BOS : Battlefield Operating System (전장기능 체계)

BP : Battle Position (전투진지)

BSC : Battle Simulation Center (전투모의센타)

BTCS : Battalion Tactical Computer System (대대전술 사격지휘체계)

C

C/S : Chief of Staff (참모장)

C4I : Command, Control, Communication, Computer and Intelligence

CA : Civil Affairs (민사)

CA : Coordination Altitude (협조고도)

CAD : Computer Aided Design (컴퓨터를 이용한 설계)

CAL : Caliber (구경)

CALO : Combined Air Lift Office (연합공수 사무소)

CAM : Computer Aided Manufacturing (컴퓨터를 이용한 제조)

CAP : Combat Air Patrol (전투공중초계)

CAS : Close Air Support (근접항공지원)

CASIC : Combined All Sources Intelligence Center (연합정보 종합상황실)

CATF : Combined Amphibious Task Force (연합상륙기동부대)

CB : Common Battery (공전식)

CBR : Chemical, Biological, Radiological (화생방)

CBU : Cluster Bomb Unit (확산탄)

CCC : Command and Control Center (지휘통제본부)

CCIR : Commander's Critical Information Requirement (지휘관 중요첩보요구)

CCP : Communications Check Point (통신확인점)

CCT : Combat Control Team (전투통제반)

CEOI : Communication Electronics Operating Instructions (통신전자운용지시)

CEWI : Combat Electronics Warfare Intelligence (전투전자전정보)

CF : Command Facility (지휘시설)

CFA : Covering Force Area (엄호부대 경계지대)

CFC : ROK/US Combined Forces Command (한미연합사령부)

CFL : Coordinated Fire Line (사격협조선)

CG : Commanding General (사령관)

CH : Cargo Helicopter (수송헬기)

CIA : Central Intelligence Agency (미 중앙정보부)

CID : Criminal Investigation Department (범죄수사단)

CIO : Chief Information Officer

CLF : Combined Landing Force (연합상륙군)

CLGP : Cannon-Launched Guided Projectile (레이저 유도포탄)

CMFC : Combined Marine Forces Command (연합 해병사령부)

COMSEC : Communication Security (통신보안)

COP : Combat Outpost (전투전초)

COP : Command Observation Post (지휘관측소)

CP : Check Point (확인점)

CP : Command Post (지휘소)

CP : Contact Point (접촉점)

CP : Control Point/Post (통제점/소)

CPMX : Command Post Maneuver Exercise (지휘조 기동연습)

CPX : Command Post Exercise (지휘조 연습)

CRAC : Combined Rear Area Coordinator (연합후방지역 조정관)

CRC : Control and Reporting Center (중앙관제소)

CRP : Control and Reporting Post (지방관제소)

CRTOC : Combined Rear Tactical Operations Center (연합후방 전술작전본부)

CS : Combat Support (전투지원)

CSCT : Combat Support Coordination Team (전투지휘협조반)

CSG : Chief of the General Staff (참모총장)

CSG : Corps Support Group (군단지원단)

CSOC : Combat Service Operations Center (전투근무 작전본부)

CSR : Controlled Supply Rate (통제보급률)

CSS : Combat Service Support (전투근무지원)

CSSOC : Combat Service Support Operation Center (전투근무지원 작전본부)

CT : Computer Tomography (컴퓨터 단층촬영)

CTB : Combined Targeting Board (연합표적위원회)

CV : Aircraft Carrier (항공모함)

DAG : Division Artillery Group (사포군)

DAO : Deep Area Operations Plan (중심지역 작전계획)

DASC : Direct Air Support Center (직접항공지원본부)

DBA : Deep Battle Area (종심전투지역)

DBSL : Deep Battle Synchronization Line (종심 전투통합선)

DCA : Defensive Counter Air (방어제공)

DCG : Depute CG (부사령관)

DD : Destroyer (구축함)

DEFCON : Defense Readiness Condition (방어준비태세)

DMZ : DeMilitarized Zone (비무장 지대)

DN : Drawing Number (도면번호)

DO : Duty Officer (당직장교)

DP : Dual Purpose (이중목적)

DS : Direct Support (직접지원)

DZ : Drop Zone (투하지역)

EA : Electronic Attack (ECM) (전자공격)

EC : Electronic Combat (전자전)

ECAS : Emergency Close Air Support (비상 근접항공지원)

EEFI : Elemetary Element of Friendly Information (우군첩보 기본요소)

EENT : End of Evening Nautical Twilight (해상박명말)

EFA : The European Fighter Aircraft (차세대 유럽형전투기)

EMERGCON : Emergency Condition (긴급/비상상황)

ENG : Engineer Joint/UNC (연합/유엔사 공병참모)

EOB : Enemy Order of Battle (적 전투서열)

EOD : Explosive Ordnance Disposal (폭발물처리반)

EP : Electronic Protection (ECCM) (전자보호)

EPW : Enemy Prisoner of War (적 포로)

ES : Electronic Warfare Support (ESM) (전자전 지원)

ESCORT : Enhancement to Sensors, Correlation Reconnaissance and Targeting
ETA : Estimate Time of Arrival (도착예정시간)
EUSA : Eighth United States Army (미 8군)
EW : Electronic Warfare (전자전)

F

FA : Field Artillery (야전포병)
FAC : Forward Air Controller (전방 항공통제관)
FARP : Forward Arming and Refuling Post (전방무장 및 연료 재보급소)
FASCAM : Family of Scatterable Mines (살포식 지뢰)
FAST : Forward Area Support Team (전방지역 지원팀)
FB : Forward Boundary (전방전투지경선)
FBI : Federal Bureau of Investigation (미 연방 수사국)
FCC : Flight Coordination Center (비행협조소)
FD EX : Focus Dragon Exercise (맹룡 연습)
FDC : Fire Direction Center (화력지휘소)
FDO : Fire Direction Officer (화력지휘장교)
FE/FDO : Force Enhancement/Flexible Deterrent Option (전투력 증강/신속억제방안)
FE : Foal Eagle Exercise (독수리연습)
FEBA : Forward Edge of Battle Area (전투지역전단)
FFA : Free Fire Area (화력자유지역)
FFIO : Field Artillery Intelligence Officer (포병정보장교)
FFIR : Intelligence Requirement (우군첩보요구)
FIDO : Fighter Duty Officer (전투비행 당직장교)
FIST : Fire Support Team (화력지원팀)
FL : Focus Lens (포커스렌즈 연습)
FLOT : Forward Line of Own Troop (전선)
FM : Field Manual (야전교범)
FM : Frequency Modulation (주파수 변조)
FMS : Foreign Military Sales (대외 군사판매)
FO : Forward Observer (관측장교)
FOP : Forward Orbit Point (전방대기지점)
FOUO : For Official Use Only (대외비)
FPF : Final Protective Fire (최후방어사격)
FRAGO : Fragmentary Order (단편명령)
FRMS : Formation Resources Management System (편성부대 자원 관리시스템)
FROG : Free Rocket Over Ground (프로그 미사일-지대지)
FS cell : Fire Support cell (통합화력지원실)
FSCC : Fire Support Coordination Center (화력지원 협조소)
FSCL : Fire Support Coordination Line (화력지원 협조선)
FSO : Fire Support Officer (화력지원장교)
FTR : Fighter (전투기)
FTX : Field Training Exercise (야외 훈련 연습)

G

GBU : Guided Bomb Unit (항공유도 폭탄류)
GCA : Ground Controled Approach (지상관제 접근)
GCC : Ground Component Command (지상구성군 사령관)
GCCS-K : Global Command and Control System Korea(주한 범세계 지휘통제체계)
GEMSS : Ground Emplaced Mine Scattering

System (지상살포식 지뢰)

GFAC : Groud Forward Air Controller (지상 전방항공통제관)

GIS : Geographic Infomation System (지형정보 체계)

GLO : Ground Liaison Officer (지상군 연락장교)

GM : Guided Missile (유도탄)

GNP : Gross National Product (국민총생산)

GOB : Ground Order of Battle (지상군 전투서열)

CONUS : Continental United States

GOP : General Outpost (일반전초)

GP : General Purpose (일반목적용)

GP : Guard Post (감시초소)

GPS : Global Position System (인공위성 항법 장치)

GS : General Support (일반지원)

GSR : General Support and Reinforcing (일반 지원 및 화력증원)

GV : Grid Variation (도북편각, 자편각, GM각)

H : Howitzer (곡사포)

HAHO : High Altitude High Open (고공 낙하산 개방)

HALO : High Altitude Low Open (고공강하)

HARM : High Speed Anti-Radiation Missile (고속 방사선 추적 유도탄)

HAWK : Homing All the Way Killer

HE : High Explosive (고폭탄)

HEAP : Hige Explosive Armorpiercing (대전차 철갑탄)

HEAT : Hige Explosive Anti Tank (대전차 고폭탄)

HEDP : High Explosive Dual Purpose (이중목적 고폭탄)

HEI : High Explosive Incendiary (고폭 소이탄)

HEI-T : High Explosive Incendiary-Tracer (고폭 소이예광탄)

HF : High Frequency (고주파)

HPT : High Payoff Target (핵심표적)

HQ : Headquarters (본부사령부)

HRP : Highway Regulation Point (육로조정소)

HSLLADS : High Speed Low Level Air Delivery System (고속저공 공중투발체계)

HUD : Head Up Display (계기판)

HVDP : Heavy Drop (중장비투하, 대량투하)

HVT : High Value Target (고가치 표적)

IAEA : International Atomic Energy Agency (국제원자력 기구)

ICAS : Immediate Close Air Support (긴급근접 항공지원)

ICBM : InterContinental Ballistic Missile (대륙간 탄도미사일)

ICM : Improved Capabilities Missile (개량미사일)

ICM : Improved Conventional Munition (개량탄)

ID : Infantry Division (보병사단)

IEW : Intelligence and Electronics Warfare (정보 및 전자전)

IFF : Identification Friend or Foe (적 · 아식별)

IGS : Inertial Guidence System (관성유도장치)

IIR : Imaging Infra-Red (적외선 영상)

ILL : Illuminating (조명탄)

IMF : International Monetary Fund (국제통화 기금)

INS : Inertial Navigation System (관성항법장치)

INT : INTerdiction (항공후방차단)

IO : Input and Output (입출력)

IOC : International Olympic Committee (국제 올림픽위원회)

IOE : Irregular Outer Edge (불규칙 지뢰지대)

IP : Initial Point (최초진입지점)

IP : Internet Protocol (인테넷 접속장치)

IPB : Intelligence Preparation of the Battle Field (전장정보분석)

IPIR : Initial Photographic Interpretation Report (최초사진판독)

IR : Infrared (적외선)

IRBM : Intermediate Range Ballistic Missile (중거리 탄도유도탄)

IROAN : Inspect Repair Only As Necessary (아이론 정비)

ISO : International Standard Organization (국제 표준화기구)

ITO : Integrated Tasking Order (통합임무명령서)

J

JAAT : Joint Air Attack Team (합동공중공격반)

JAM : Jamming (전파방해)

JCS : Joint Chiefs of Staff (합동참모본부)

JRAC : Joint Rear Area Coordinator (합동 후방지역조정관)

JSA : Joint Security Area (공동경비구역)

JSEAD : Joint Suppression of Enemy Air Defence (합동 적방공제압)

JSTARS : Joint Surveillance and Target Attack Radar System (합동감시 표적공격 레이다 체제)

JTF : Joint Task Force (합동기동부대)

JUSMAG-K : Joint United States Military Affairs Group to Korea (주한미합동군사업무단)

K

KACC : Korean Airspace Control Center (한국 공역통제본부)

KADIZ : Korea Air Defence Identification Zone (한국방공식별구역)

KAIST : Korea Advanced Institute for Science and Technology (한국과학기술원)

KALCC : Korea Air Lift Control Center (한국 공수통제본부)

KAME : Korean Air Mobility Element (한국 공중기동반)

KAMO : Korea Airlift Management Office (한국 공수관리소)

KATUSA : Korean Augmentation To United States Army (미육군 배속 한국군)

KCOIC : Korean Combat Operations Intelligence Center (한국 전투작전정보본부)

KCRCC : Korean Combined Rescue Coordination Center (한국 연합구조협조본부)

KEDO : Korean Energy Development Organization (한반도 에너지개발기구)

KFP : Korea Fighter Program (한국군 차세대 전투기사업)

KIST : Korea Institute of Science and Technology (한국과학기술원)

KMAC : Korea Army Mapping Center (한국육군지도창)

KNR : Korea National Railroad (한국철도청)

KS : Korea (Industrial) Standards (한국표준)

L

LA : Live Ammunition (실탄)
LAN : Local Area Network (근거리통신망)
LANTIFN : Low Altitude Navigation Targeting Infrared for Night
LAPES : Low Altitude Parachute Extraction System (저공낙하산 추출장치)
LATN : Low Altitude Tactical Navigation (저고도 전술비행)
LAU : Launcher Unit (발사대)
LAW : Light Antitank Weapon (경 대전차화기)
LB : Local Battery (자석식)
LC : Line of Contact (접촉선)
LCM : Landing Craft Medium (상륙정)
LCU : Landing Craft Utility (보급상륙정)
LD : Line of Departure (출발선/공격개시선)
LF : Landing Force (상륙부대)
LGB : Laser Guided Bomb (레이저 유도폭탄)
LGM : Laser Guided Missile (레이저 유도미사일)
LMG : Light Machine Gun (경기관총)
LNG : Liquefied Natural Gas (액화천연 가스)
LO : Liaison Officer (연락장교)
LOC : Lines of Communication(Logistic routes) (병참선)
LP : Listening Point (청음초)
LP : Loading Point (적재지점)
LPG : Liquefied Petroleum Gas (액화석유가스)
LPG : Liquefied Propane Gas (액화프로판가스)
LRBM : Long Range Ballistic Missile (장거리 탄도 미사일)
LST : Landing Ship Tanks (전차상륙함)
LZ : Landing Zone (착륙지역)

M

M-day : Mobilization-day (동원개시일)
M/F : Maintenance Float (정비대충)
MARV : Maneuverable Reently Vehicle (핵탄두 유도탄)
MASP : Mobil Ammunition Supply Point (가동탄약보급소)
MBA : Main Battle Area (주전투지역)
MBT : Main Battle Tank (전투용 주전차)
MC : Medical Command (의무사령부)
MCA : Movement Control Agency (이동관리대)
MCM : Military Committee Meeting (군사위원회 회의)
MCRC : Master Control and Reporting Center (중앙방공통제소)
MDL : Military Demarcation Line (군사분계선)
METT+TC : Mission, Enemy, Terrain, Troops and Time available, Civil affairs
MI : Military Intelligence (군사정보)
MICR : Magnetic Ink Character Reader (자기잉크 문자판독기)
MLRS : Mutiple Launch Rocket System (다련장 로겟 체계)
MOPP : Mission Oriented Protection Posture (임무형 보호태세)
MOS : Military Occupational Specialty (군사주특기)
MOT : Mortar (박격포)
MP : Military Police (헌병)
MRBM : Mid Range Ballistic Missile (중거리탄도미사일)
MRE : Meals Ready to Eat (전투식량)
MRL : Multiple Rocket Launcher (다련장 로켓 발사대)

MRR : Minimum Risk Route (최소위험경로)

MSD : Minimum Safe Distance (최소안전거리)

MSK : Mun Shu Ko Bunker

M&A : Merger and Acquisition (기업합병)

N

NAD : Not on Active Duty (예비역)

NAI : Named Area of Interest (중요감시지역)

NASA : National Aeronautics and Space Administration (미 항공우주국)

NATO : North Atlantic Treaty Organization (북대서양 조약기구)

NBC : Nuclear, Biological, Chemical (화생방)

NCC : Naval Component Command (해군구성군 사령부)

NCO : Noncommissioned Officer (부사관)

NFA : No Fire Area (화력금지구역)

NFL : No Fly Line (비행금지선)

NOD : Night Observation Device (야관관측장비)

NOS : Network Operating System (통신망 운영체계)

NP : Nepalm Incendiary (네이팜소이탄)

NPT : Nuclear Non-Proliferation Treaty (핵 확산금지 조약)

NVG : Night Vision Goggle (야시장비)

NVS : Night Vision Sight (야간관측기)

O

OA : Office Automation (사무자동화)

OAC : Officer's Advanced Course (고등군사반)

OB : Order of Battle (전투서열)

OBC : Officers' Basic Course (초등군사반)

OCA : Offensive Counter Air (공세제공)

OCR : Optical Character Reader (광학 문자판독기)

OECD : Organization for Economic Cooperation and Development (경제협력 및 개발기구)

OMG : Operational Maneuver Group (작전기동단)

OMR : Optical Mark Reader (광학 표시판독기)

OP : Observation Post (관측소)

OP : Orbit Point (선회대기지점)

OPCON : Operational Controls (작전통제)

OPLAN : Operation Plan (작전계획)

OPMS : Officer Personnel Management System (장교 인사관리체계)

OPORD : Operation Order (작전명령)

OPSEC : Operation Security (작전보안)

OT : Observer Target (관목선)

OVM : On Vehicle Material (차량부수기재)

P

PC : Personnel Command (인사사령부)

PCS : Permanent Change of Station

PGM : Precision Guided Munition (정밀유도폭탄)

PIR : Priority Intelligence Requirements (우선정보요구)

PKO : Peace Keeping Operations (유엔 평화유지활동)

PL : Phase Line (통제선)

PLADS : Parachute Low Altitude Delivery System (저고도 낙하산 투발체계)

PLL : Prescribed Load List (규정휴대량 목록)

PM : Pulse Modulation (펄스변조)

PR : Public Relation (공공관계/홍보)

PRC : Portable Radio Communication (휴대용 무선 통화장비)

Pre-ITO : Prepositioned Integrated Tasking Order (기계획 통합임무명령서)

PRI : Preliminary Rifle Instruction (사격술예비 훈련)

PT : Physical Training (체력단련)

PZ : Pickup Zone (인수지대, 이륙지대)

R

RAAM : Remote Anti Armor Mine (원격조정 대전차지뢰)

RADAR : Radio Detection And Ranging (레이다)

RAOC : Rear Area Operations Center (후방지역작전본부)

RAP : Rocket Assisted Projectile (로켓보조탄두)

RAPCON : Radar Approach Control (레이다 접근관제)

RCT : Regimental Combat Team (연대전투단)

RDO : Reconnaissance Duty Officer (정찰담당 장교)

RECCE : Reconnaissance (정찰)

RF : Reinforcing (화력증원)

RFA : Restrictive Fire Area (화력제한지역)

RFL : Restrictive Fire Line (사격제한선)

RISTA : Reconnaissance, Intelligence, Surveillance and Target Acquisition

RJ : Road Junction (교차점)

ROA/ROZ : Restricted Operations Area/Zone (작전제한구역/지대)

ROE : Rules Of Engagement (교전규칙)

ROP : Rear Orbit Point (후방 선회대기지점)

ROTC : Reserve Officer's Training Corps (학군단)

RP : Release Point (분진점)

RP : Rendezvous Point (집결지점)

RPV : Remotely Piloted Vehicle (원격조정 무인정찰기)

RSOI&FE : Reception, Staging, Onward movement, Integration and Foal Eagle(전시증원 및 독수리연습)

RSOI : Reception, Staging, Onward movement and Intergration (수용, 대기, 전방이동및통합)

RSR : Required Supply Rate (소요보급률)

RTB : Return To Base (기지귀환)

S

SACC : Supporting Arms Coordination Center (지원화기 협조본부)

SAH : Semi-Active Homing (반능동 유도장치)

SALT : Strategic Arms Limitation Talk (전략무기제한 협정)

SAM : Surface to Air Missile (지대공미사일)

SCM : Security Cooperative Meeting (한미 안보협조회의)

SCUD : Medium/Long Range Ballistic Missile(NK)

SDI : Strategic Defence Initiative (전략방위구상)

SEAD : Suppression of Enemy Air Defence (적방공제압)

SF : Special Forces(Ranger/Commando) (특수부대)

SITREP : Situation Report (상황보고)

SLBM : Sea Launched Ballistic Missile (함상발사 탄도유도탄)

SLBM : Submarine Launched Ballistic Missile (잠수함발사 탄도유도탄)

SLCM : Sea Launched Cruise Missile (함상발사 순항유도탄)

SLOC : Sea Lines Of Communications (해상병참선)

SN : Serial Number (일련번호)

SOCKOR : Special Operations Command Korea (주한미 특전사령부)

SODO : Senior Operations Duty Officer (선임작전 당직장교)

SOI : Signal Operation Instructions (통신규정)

SOP : Standing Operation Procedure (부대예규)

SP : Start Point (출발점)

SQ : Squadron (비행대대)

SRAGM : Short Range Ait to Ground Missile (단거리 공대지 유도탄)

SRAM : Short Range Attack Missile (단거리 공격용 유도탄)

SRBM : Short Range Ballistic Missile (단거리 탄도유도탄)

SRH : Semiactive Radar Homing (반능동 레이다추적)

SSBM : Strategic Submarine Ballistic Missile (전략잠수함 탄도미사일)

SSM : Surface to Surface Missile (지대지 미사일)

STAFF : Smart Top Attack Fire and Forget (비유도 스마트탄)

STOL : Short Take Off and Landing (단거리 이착륙)

SROKA headquarters=Second Republic of Korea Army (제2군 사령부)

T

TA : Table of Allowance (할당표)

TA : Table of Authorization (인가표)

TAACOM : Theater Army Area Command (전구 육군 지역사령부)

TACAN : TACtical Air Navigation (전술항법 보조장치)

TACC : Theater Air Control Center (전구 항공통제본부)

TACP : Tactical Air Control Party (전술항공통제반)

TACS : Theater Air Control System (전구 항공통제체제)

TAI : Target Area of Interest (관심표적지역)

TAL : Tactical Action Line (전술조치선)

TAL : Theater Air Lift (전구공수)

TALO : Tactical AirLift Liaison Officer (전술공수 연락장교)

TAR : Tactical Air Reconnaissance (전술항공정찰)

TASE : Tactical Air Support Element (전술항공지원반)

TCP : Traffic Control Point (교통통제소)

TD : Table of Distribution (분배표)

TE : Table of Equipment (장비인가표)

TF : Task Force (특수임무부대)

TG : Tear Gas (최루탄)

TGT : Target (표적)

TM : Technical Manual (기술교범)

TMD : Theater Missile Defense (전구 미사일 방어)

TMO : Traffic Management Office(r) (교통관리소)

TNT : Trinitrotoluence

TO : Table of Organization (편성표)

TOD : Thermal Observation Device (열상관측장비)

TOD : Time Of Departure (출발시간)

TOT : Time On Taget(Artillery Support) or Time Over Taget(Air Support)

TOW : Tube−launched , Optically tracked, Wire guided Missile

TP : Target Practice (표적연습용)

TPFDO : Time-Phased Force Deployment Data (시차별 부대전개제원/증원전력)

TS : Team Spirit (팀스피리트 훈련)

U

UAV : Unmanned Aerial Vehicle (무인항공기)

UFL : Ulchi-Focus Lense (을지포커스렌즈 연습)

UH : Utility Helicopter (다목적용 헬기)

UHF : Ultra High Frequence (극 초단파)

UNC : United Nations Command (유엔군 사령부)

UNICEF : United Nations Children's Fund (유엔 아동기금)

UNO : United Nations Organization (국제연합기구)

USAFK : United States Air Forces, Korea (주한 미공군)

USFK : United States Forces, Korea (주한 미군)

USMFK : US Marine Forces, Korea (주한 미해병대)

USNFK : US Naval Forces, Korea (주한 미해군)

UTM : Universal Transverse Mercator (만국횡단 마케이터)

V

VCR : Video Cassette Recorder (비디오 녹화기)

VHF : Very High Frequence (초단파)

VIP : Very Important Person (귀빈)

VSTOL : Vertical and Short Take Off Landing (수직 및 단거리 이착륙)

VT : Variable Time fuse (가변시한신관)

VTOL : Vertical Take Off and Landing (수직이착륙)

VUL : Vulcan (발칸포)

W

WEI : Weapon Effectiveness Index (무기효과지수)

WG : Wing (비행단)

WGS : Worldwide Geological System (범세계지도체계)

WHNS : Wartime Host Nation Support (전시지원)

WHO : World Health Organization (세계보건기구)

WOC : Wing Operations Center (비행단 작전본부)

WP : White Phosphorous (백린탄)

WRSA : War Reserved Stock For Allies (동맹군을 위한 전시 예비 비축물자)

WS : Weapons System (무기체계)

WTO : Warsaw Treaty Organization (바르샤바조약기구)

X

X : Experimental (실험용)

XINT : Air Alert Interdition (공중대기 후방차단)

XO : Executive Officer (보좌관, 부지휘관)

군사영어

초　판 1쇄 인쇄 | 2013년 3월 20일
초　판 1쇄 발행 | 2013년 3월 25일
개정판 1쇄 발행 | 2015년 3월 6일
개정판 4쇄 발행 | 2020년 11월 9일

저자　한종철
발행처　형설출판사
경기도 파주시 회동길 37-23 · **전화** (031) 955-2361~4 · **팩시밀리** (031) 955-2341
발행인　장진혁
등록　라-제9호 · 1962년 5월 1일
홈페이지　http://www.hyungseul.co.kr
e-mail　hs@hyungseul.co.kr

정가 16,000원

ISBN 978-89-472-7598-9 93740

이 도서의 국립중앙도서관 출판시도서목록(CIP)은 서지정보유통지원시스템 홈페이지(http://seoji.nl.go.kr)와 국가자료공동목록시스템(http://www.nl.go.kr/kolisnet)에서 이용하실 수 있습니다.(CIP제어번호 : CIP2015005572)

MEMO